James Monteith

Monteith's physical and political Geography

James Monteith

Monteith's physical and political Geography

ISBN/EAN: 9783337132767

Printed in Europe, USA, Canada, Australia, Japan

Cover: Foto ©ninafisch / pixelio.de

More available books at **www.hansebooks.com**

NATIONAL GEOGRAPHICAL SERIES.

MONTEITH'S

PHYSICAL AND POLITICAL

GEOGRAPHY;

IN TWO PARTS.

PART I.

GEOGRAPHY TAUGHT AS A SCIENCE;

WRITTEN AND ILLUSTRATED ON THE PLAN OF OBJECT TEACHING.

PART II.

LOCAL AND CIVIL GEOGRAPHY

CONTAINING

MAPS REMARKABLE FOR THEIR CLEARNESS, AN IMPROVED SYSTEM OF MAP EXERCISES AND OF MAP DRAWING, POLITICAL AND ASTRONOMICAL GEOGRAPHY, AND A PRONOUNCING VOCABULARY OF GEOGRAPHICAL NAMES.

BY JAMES MONTEITH,

Author of a Series of School Geographies and Wall Maps, and a Juvenile History of the United States.

A. S. BARNES AND COMPANY,

NEW YORK, CHICAGO, AND NEW ORLEANS.

Character of the Work.

PART I.

Its Style.—In the preparation of this portion of the work, the author has sought to present the subject of Geography *as a Science;* and, at the same time, in a style calculated to attract and interest the pupil. Avoiding the use of all technical terms that would perplex the young learner, he has endeavored to explain its leading principles by means of familiar language and comparisons, and of suggestive illustrations, according to the *Object System* of instruction. For example, on page 19, the theory of volcanic action is explained by reference to a cake which is burst open at the top, the heat of the oven representing the heat of the earth's interior.

Again; on page 27, boiling springs, such as the geysers, are illustrated by means of a tea-kettle.

The Text is divided into Short Paragraphs so constructed that the commencement of each appears in prominent type and readily suggests the subject and the questions.

It Teaches:—That the earth was formed to be the temporary dwelling-place of mankind; and to that end, were created the land, with its mountains and plains; the water, with its mighty ocean and its running brooks; besides air, light, heat, plants, and living creatures:

That all the objects which we behold, whether organic or inorganic, whether on the surface or below the surface, with all the various phenomena of the earth, exert important influences upon each other and work together in harmony for the well-being of the human race.

Effect upon the Mind of the Pupil.—Throughout the work the aim has been not only to impart valuable information, but especially to cultivate the learner's powers of observation and reasoning; and, as he views the wonders, the beauty, and the perfection of Nature's works, his thoughts are thereby directed to the Creator, by whom all things were made and adapted to the development of human life and happiness.

The Index and General Review serve both as a Reference and as a system of General Exercises.

Among the Works on Geology and Geography which have been consulted by the author, are those of Lyell, Hitchcock, Dana, Miller, Johnston, Milner, and Ritter.

PART II.

The Maps have been executed with reference to clearness and freedom from detail.

The Largest City in each State or country appears in large capital letters. The capital is designated by a *. For the names of places represented on the maps by numbers, see Appendix.

Accompanying the Maps of the United States, are Additional Exercises adapted for use in each State separately.

The Principal Railroads are shown by finely dotted lines, and in connection with them are questions on "Routes of Travel."

In the "Reviews" are given the population of the largest cities, the height of the highest mountains, and the length of the largest rivers. The "General Review" contains questions promiscuously arranged.

The Political Geography, fully illustrated, gives a concise view of the leading features of the Countries and States,—their topography, soil, climate, productions, &c.

Comparative Sizes and Latitudes of Countries, States and Cities are shown on the margins of the Maps. (See also p. 102.)

PART I.

GEOGRAPHY.

CONTENTS.

LIST OF MAPS.

INTRODUCTORY.

1. The robin builds her nest in the tree for the *Purpose* of there depositing her eggs, and of bringing forth and protecting her young.

2. For the *Purpose* of protection and comfort men build houses, found cities, and establish governments. *Purpose, therefore, leads to Design and Action.*

3. When you look at a beautiful house, and observe the peculiar fitness of the various parts to each other, you are certain that it was made for the security and enjoyment of the family within; and that the workmen shaped and placed the materials under the direction of an intelligent architect, who *Formed the Plan before the Work was commenced.*

4. So, when you look abroad, you see a beautiful world, which was made for the enjoyment and benefit of the whole human family.

5. *Man could not exist without Food;* therefore the earth yields her manifold productions of grain, fruit, and vegetables, while animals, birds, and fish, also, are given for his nourishment and use. *Neither could he live without Drink;* so the earth is abundantly supplied with refreshing springs. *For Clothing* he goes to the cotton plant, the sheep, and the silkworm; from the forests and the ground he obtains all the materials for building purposes.

6. *Animal Life receives its Sustenance* from plants; *Plants receive theirs* from the soil and moisture; *Soil proceeded* originally from the hard rock; *Moisture and Clouds*, from the ocean.

7. The earth has its continents and oceans, its mountains and plains, its rocks, rains, snows, springs, and stream work harmoniously for the welfare and happiness of mankind.

8. You may conclude, then, that the whole earth, of which all these things are but parts, was made for a *Great P* by a Being of infinite wisdom, goodness, and power, according to a design formed before the beginning of the world this purpose was *to provide an Abode for Man, whose Delight would be to praise, honour, and serve Him.*

Section II.

CREATION OF THE EARTH.

The Earth's Surface covered with Water.

1. The *Growth of a Plant* progresses slowly and systematically; from the seed comes a stem, then leaves, blossoms, and fruit. So was the process by which the world was made from chaos,—slow, gradual, and in accordance with the provisions of a well-ordered plan established by Divine wisdom.

2. The *Earth's Formation from Chaos* may be illustrated by an egg, whose fluid substances, by a certain application of heat, and in a certain time, are changed into a beautiful, living bird.

3. "*In the Beginning,* God created the heaven and the earth." In time the earth received its globular shape, and consisted of a heated, earthy matter in a fluid form, the outside of which, becoming cool and hard, formed a kind of crust around the mass. Entirely surrounding this crust was water, and surrounding the water was the atmosphere, containing dark, heavy clouds.

4. The *Rain formed no Springs,* watered no fields. It fell only upon the salt ocean, for the whole outer side of the earth's crust constituted the bed of the ocean.

"And God said, Let the Waters under the Heaven be gathered together, and let the Dry Land appear."

5. *By Convulsions within the Earth,* parts of the crust were forced upward through the water, and became dry land.

6. The *Land first Raised* consisted only of masses of hard rock, on which no tree or plant could grow.

7. *There was no Soil* until the rock was broken and pulverized by the action of the waves, air, rain, heat, and cold.

8. From the grinding together of fragments of the rock, came stones, pebbles, gravel, and sand.

9. "And God called the dry land Earth, and the gathering together of the waters called He Seas."

"And the Earth brought forth Grass, and Herb yielding Seed after his kind, and the Tree yielding Fruit."

10. The *Violent Agitation of the Earth's Interior* greatly disturbed the bed of the ocean, causing the depression of some parts and the elevation of others; in the former, the sea became deeper, and in the latter, more shallow.

11. *Portions of the Ocean's Bed* were in this way brought up to the surface, then above it; and, covered with the pulverized or disintegrated rock which had long been settling upon them, these tracts of land, in time, supported trees and plants which received their nourishment from both the soil and the atmosphere.

"And God created great Whales, and every Living Creature that moveth, which the Waters brought forth abundantly, after their kind, and every Winged Fowl after his kind."

"And God made the Beast of the Earth after his kind and Cattle after their kind, and every thing that creepeth upon the Earth after his kind."

12. The *Various Species of Animals* which have lived upon the earth were not all created at once.

13. The *Lower Orders* came first; and, as centuries rolled on, other and superior classes of animals came successively into existence.

14. *Insects, Fish, and Reptiles* were created before the horse or the ox; and all species of animals were created before Man.

15. *With Plants, also, this was the case.* The first vegetation consisted of sea-weeds; then, with the improvement of the soil, new and superior varieties of plants and trees appeared.

16. *These facts have been ascertained* from investigations below the earth's surface, where the forms or remains of plants and animals, which lived in successive periods, are found in the order of their creation; those created first being farthest below the surface.

17. *We see the Law of Gradual Development* exemplified in the growth of the trees and living creatures; geologists observe it, also, in the rocks and sands of the earth.

"And God said, Let us make Man in our Image, after our Likeness; and let them have dominion over the Fish of the Sea, and over the Fowl of the Air, and over the Cattle and over all the Earth, and over every Creeping Thing that creepeth upon the Earth."
"So God created Man in his own Image."

18. The general order of Creation was as follows:

(1.) Chaos.

(2.) Melted Matter in the form of a globe.

(3.) The Globe composed of melted matter having a crust, which was entirely surrounded by water.

(4.) Parts of the Crust upheaved through the sea, forming dry land.

(5.) Pulverized Rock; forming soil.

(6.) Land alternately upheaved and submerged.

(7.) Vegetation.

(8.) Animal Life.

(9.) Man.

19. The *Observing Pupil has now Learned* two important facts; first, that God made the world, with all it contains, not at once, but step by step, on a wise and definite plan; second, that He made it for the use of man.

20. *For the Life and Happiness of Mankind* there are provided, not only the objects and creatures mentioned in the beginning of Genesis, but also numberless features and phenomena of the earth, such as its atmosphere, climates, currents, rain, mountains, plains, and productions.

21. The *Science of Geography* properly embraces an investigation into the laws which control the conditions, changes, and phenomena in nature, as affecting the life and conditions of mankind.

22. Although the various departments of Geographical Science will be presented in this work in a classified form, yet it is highly important that the learner keep constantly in mind their dependence and influence upon each other; this renders repetition, to some extent, essential.

23. When considering the position and height of a chain of mountains, the course of the winds, or of an ocean current, he should observe the influences exerted by each upon climate, vegetation, and the pursuits of man in different regions.

24. These *Differences or Contrasts* furnish each section with its own characteristic productions, and lead men to establish a system of trade or commerce between the nations of the earth, thus increasing their industry and wealth, furnishing incentives for exploration, and securing the civilization and enlightenment of the race.

25. The *Pupil should know*, not only that the Gulf Stream has a north-easterly direction, but also that its warmth tempers the climate of the greater part of Europe, and sheds its genial influence upon the atmosphere, productions, and inhabitants of that Grand Division. He should observe that the highest mountains are in the hot regions of the earth, where their lofty peaks, continually wrapped in snow, are faithful refrigerators, reducing the temperature of the air on the heated plains below.

26. The text, generally, is written without set questions; leaving the teacher to frame or vary them as he may wish. Interrogations, however, are made which can be answered, not directly from the text, but from the illustrations, or by inference on the part of the learner.

27. This plan cannot fail to lead youthful minds to habits of observation and reasoning, and to direct their thoughts to the wisdom, power, and goodness of the Creator.

Section III.

THE CRUST OF THE EARTH.

1. The *Crust of the Earth is the result of* the cooling of the melted matter at the surface. It becomes thicker, as ice does, by additions to its under side.

2. Scientific investigations show that the ground is affected by the sun's heat to the depth of about 50 feet; below that, the heat of the earth's interior increases according to the depth.

3. The *Internal Heat* does not extend to the surface of the earth, except on occasions of earthquakes and volcanic eruptions.

A View within the Earth's Crust.—Salt Mines of Austria, 500 Feet Deep.

4. The average increase of temperature, below where it is affected by heat from the sun, is about one degree for every 50 or 60 feet in depth; accordingly, at the depth of about 50 miles, the heat would be sufficient to melt all known rocks.

5. Geologists have variously estimated *the thickness of the earth's crust* to be from 20 to 200 miles.

6. Man has penetrated the earth to the depth of about one mile.

7. The *Crust*, if 20 miles in thickness, bears the same proportion to the whole earth that an egg-shell does to the egg.

8. The *Height of the Highest Mountains* in the world is about 5 miles, yet the distance from the level of the sea to the center of the earth is 800 times greater than that.

9. The *Material* of which the earth's crust is composed is termed Rock, whether it be hard and compact, or soft and loose: it is constantly undergoing change, owing, chiefly, to the agency of air, water, and heat.

10. *Aqueous Rocks* are those formed by the agency of water. They consist of the sediment which has become hardened in layers or beds, and are called Stratified.

11. *Igneous Rocks* are those formed by the agency of fire. They consist of hard, irregular masses, and are therefore called Unstratified.

A. Stratified Rock; B. Unstratified Rock; C. Melted Matter of the Earth's Interior.

12. *As the Surface*, at an early period, was entirely covered with water, where would you find the Aqueous or Stratified formations?

13. The *Igneous or Unstratified* rock found at the earth's surface has been forced up through the aqueous or stratified formations by volcanic action.

14. *In some Rocks are found* forms of animals and vegetables petrified or hardened like stone, caused, chiefly, by chemical action in nature.

15. *Geologists Show* that the greater part of the soil or mold on the earth's surface is composed of what in former ages constituted the bodies of animals, trees, and plants, mixed with mineral substances, all of which settled at the bottom of the water.

16. The petrified forms of animals and plants are called *Fossils;* the strata in which they are found are called *Fossiliferous.*

17. The *Direction of the Strata* or layers would be horizontal and parallel to each other, but for the disturbing forces of the earth's interior, which have raised the strata in parts, giving them uneven or inclined positions.

18. Where the strata are horizontal, which of them was the most recently formed? Which was first formed? What can you say of the heat of the earth's surface? Of the earth's interior? What can you say of the material which forms the earth's surface? What is the difference between aqueous and igneous rocks?

19. *Each Stratum of Hard Rock is Composed* of what had been soft mud, loose gravel, shells, vegetable and animal bodies.

20. The *Forms of Animal Bodies* in one stratum have been found to differ from those in the stratum below or above it, proving that at successive periods there lived successive species of animals.

The Form and Surface of the Earth.

Section IV.

THE FORM OF THE EARTH.

1. The *Form of the Earth* is that of a "*Globe*," or "*Sphere.*" For this reason the topmast of a ship approaching us is first seen, then the sails, and, lastly, the body of the ship.

2. If you look around when at sea, or on a plain, what kind of a line limits your view? What is the name of that circle?

3. If you sail or move from one place to another, does your horizon change? If you go to the top of a mountain, or any eminence, how is the extent of your horizon affected?

4. *Who can see an Approaching Ship first*, the man at the foot, or the one at the top of a mountain? Which has the more extended horizon?

5. Which of these two men can first see the sun rise in the morning? Sun set? Is the day longer to one than to the other? To whom? Why does the light on a distant light-house appear to be on the surface of the water?

6. The *Continents, Islands, and Mountains* which we now behold were not formed at once; some parts were raised suddenly, but most of the land elevations were the work of ages.

7. The *Inequalities of the Earth's Surface* are no greater, relatively, than the roughness on the surface of an orange; and, although appearing to the careless observer as accidental and meaningless, they exert, nevertheless, important influences upon the conditions of mankind, and are in accordance with the wise designs of the Creator.

8. *One-fourth of the Earth's Surface* is land; three-fourths, water. In other words, the internal forces have thus far caused the elevation of one-fourth of the ocean's bed.

9. As the *Bed of the Ocean along the Coasts* is inclined, what would be the effect of an increase in the volume of water upon the size of continents and islands? Upon their elevations? What would be the effect upon the same if the volume of water should be diminished? What, if the ocean's bed should be suddenly depressed? Elevated?

10. The *Bed of the Ocean comprises* the greatest depressions of the earth's crust; and, in its unevenness, it is like the land above the water level.

11. The *Ocean acts an Essential Part* in the unfolding of the Creator's design to benefit mankind. It is not only the highway between the nations of the earth, but also the modifier of climate, and the vast reservoir whence the land receives its entire supply of water for the support of all life, whether animal or vegetable.

12. *If the Ocean covered the whole Surface* of the earth, could man exist?

13. *If the Surface consisted entirely of Land*, the absence of water would forbid the existence of mankind; for all vapor, clouds, rain, springs, streams, and lakes would disappear. All the fresh water of the land is raised from the great reservoir, the ocean, by the combined agencies of the sun and air, acting like a great pump and sprinkler.

14. *At the Earth's Surface* there are in contact three elements,—water, land, and air; to deprive man of any one of these would be to deprive him of life.

15. The *Earth covered with Land and Water*, but without the atmosphere, could not be the abode of man, for there would be no water to drink, no air to breathe; the land, not watered by dews and rain, could not yield him food.

16. Therefore, *Two Indispensable Agents* are provided,—the sun and atmosphere.

The *Sun by his Powerful Light and Heat* so acts upon the sea that thin, fresh water called vapor is separated from it. The vapor, like a feather loosened from a bird, is borne upward by the atmosphere, and carried far away by the winds.

Vapor becomes Clouds, and afterward returns to the earth in the form of rain, dew, or snow, to water and fertilize the soil, and to scatter all over the land innumerable springs, streams, and lakes of delicious water.

17. It is evident, then, that *All Parts of the Earth*, above and below its surface, are made to harmonize and coöperate with each other as an organized whole, for the great object of the gradual perfection of the human race.

If there were no ocean, would there be any rivers or springs? Any rain or clouds?

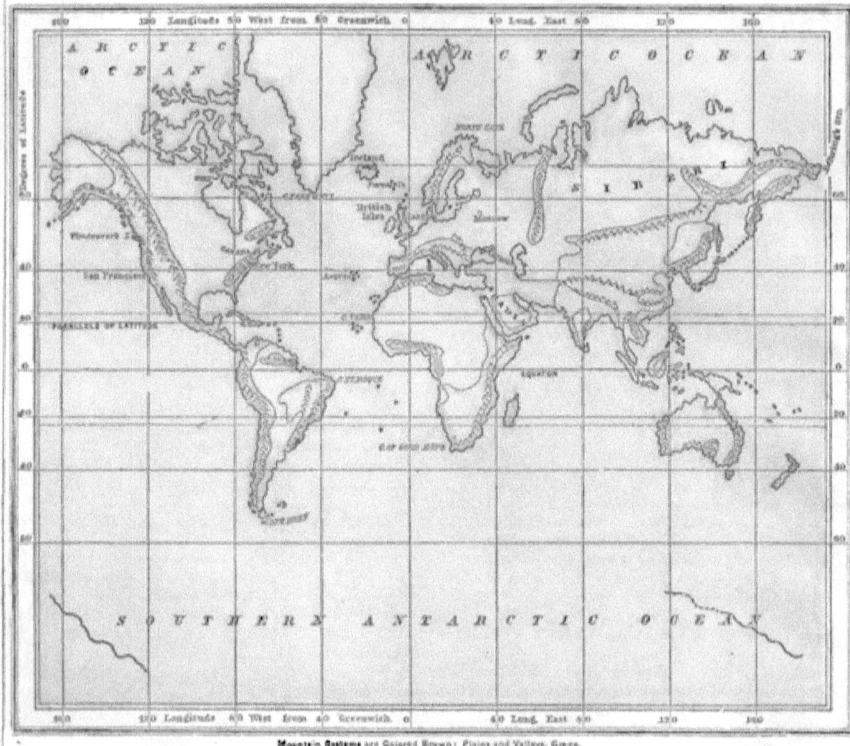

Mountain Systems are Colored Brown; Plains and Valleys, Green.

Section V.

THE CONTINENTS,—THEIR FORM.

1. The *Land on the Earth's Surface* is known, generally, as continents and islands; the continents are two in number; the Eastern or Oriental, called the Old World, and the Western or Occidental, called the New World; the islands are numerous. Australia is sometimes called a continent.

2. *When Land first emerged* from the water and came into contact with the atmosphere, it was not then as it is now, either in extent or form.

3. *None of those Large Bodies of Land* appeared, whose shapes we now trace on the globe or map; but, comparatively small points were projected, which gradually rose higher and extended more widely, according to the pressure of the forces beneath.

4. *A Continent* is entirely surrounded by water.

5. *A Continent*, with its peninsulas, highlands, lowlands, lakes, and rivers, is like a great tree that has grown from a small shrub.

6. *What is now a Vast Continent* was, at a remote period, entirely below the level of the sea; its general shape was the same then as it is now.

7. *A Continent was not raised at once*, but slowly; appearing above the water in parts.

8. *These Parts, after remaining at the Surface* for many centuries, were again submerged, and their great masses of vegetation,—trees, shrubs, and plants,—became covered over with gravel and sand.

9. *At the End of another Long Period*, the submerged vegetation and the over-lying beds would be again raised, only to undergo a similar process.

10. *Such Operations occurred* long before the creation of

man; and although to the uninformed they appear without purpose or use, they have, nevertheless, successfully contributed toward the unfolding of God's wise design in his preparation of the earth for the abode of the human race. These vegetable masses are now the exhaustless beds of coal which supply indispensable aid to the industry and comfort of man.

11. The *Wisdom of this Plan* is further recognised in the fact that coal is found, mainly, in those parts of the earth that are best fitted for human habitation;—in the United States, Great Britain, Western Europe,* British America, and China.

The Parts of the Map shown in White represent the First Land of the United States.
The Parts in Dark Shading along the Coasts remained under Water until a more recent Period.
The Dark Shading Inland were vast Tracts of Marsh and Woodland, but now they are the great Coal Fields of the Country.

12. The *Extended Lines of Elevation* which we call mountain chains or ranges seem to constitute the frame of the continents.

13. The *Slopes, Plains, and Valleys* have been shaped and fertilized by slides of great ice formations of former ages, and by frequent rains, which have washed down the dissolved and pulverized rocks, and the long decayed vegetable and animal substances; mixing them all together in a rich compound called mold, which supports the vegetation of the earth.

14. The *Great Body of Land Surface* is north of the Equator, both in the Old and in the New World, and comprises the whole of Asia, Europe, North America, Northern and Central Africa, and the northern part of S. America. South of the Equator are only three considerable tracts of land; the central and southern parts of South America, the southern part of Africa, and the island of Australia.

15. The *Land of the Two Continents* not only lies chiefly in the Northern Hemisphere, but it also widens toward the north, and narrows into peninsulas at the south, these peninsulas, also, terminating in capes pointing southward, thus giving each continent the appearance of a triangle with the apex toward the south.

16. *This Peculiar Feature makes it appear* as if the water of the ocean had originally issued in great currents from the region of the Southern Ocean, as a center, and washed away the land until arrested by the mountains and highlands of the Northern Hemisphere.

The General Form of each of the Land Divisions is that of a Triangle, the Apex pointing toward the South.

17. *Upon the Western Continent* the water seems to have encroached from the south and south-west to the foot of the vast mountain ranges which line its coast; upon Africa to the Kong and Snow Mts., and the highlands intervening; upon Asia to the Himalaya and the Ghauts Mts.

18. *With New Zealand as a Center*, describe a great circle upon the globe, dividing it into hemispheres; one will contain nearly all the land on the earth's surface, while the other will be composed almost entirely of water. These are known as the Land and Water Hemispheres. At or near the center of the Land Hemisphere are the British Isles. (See Map, p. 12.)

19. *By means of the Winds and Waves* new coasts have been formed, and others washed down to the ocean's bed; loose sand on some sea-shores is carried inland, forming drift-sand hills, such as those on the southern shore of Long Island and the eastern shore of New Jersey. In some places, these movements of the sand have been attended with destructive effects, by covering houses, farms, and villages.

Tower of a Buried Church on the East Coast of England.

20. *An Increase of the Volume of Water* would be followed by an overflowing of the land, beginning with the lowlands: thus effecting entire changes in the sizes and forms of continents.

21. The *Eastern Continent comprises* Europe, Asia, and Africa; the *Western*, North and South America.

22. The *Eastern Continent extends* in an easterly and westerly direction. Its great mountain system, commencing at Behring Strait and the Pacific Ocean, runs through central and southern Asia, and along the north and south sides of the Mediterranean Sea to Portugal in Europe, and to Morocco in Africa.

23. *These Mountains are included*, chiefly, between the parallels of 25° and 50° north latitude.

24. The *Western Continent takes its direction* from its great mountain system, which reaches from the Northern to the Southern Ocean in a north-westerly and south-easterly direction. Each of these two mountain systems is like the backbone, which gives position and strength to an animal body.

25. The *Principal Sections* of the mountain system on the Eastern Continent are the Himalay'a, Altai (*ahl-ti'*), and Stanavoy ranges of Asia; the Cau'casus, Carpathian, Alps, and Pyrenees of Europe, and the Atlas Mountains of Africa.

26. The *Sections of the great Mountain System* of the Western Continent are the Andes of South America, and the Rocky, Sierra Madre (*se-er'rah mah'dray*), Sierra Nevada (*nay-vah'dah*), and Cascade of North America. These great ranges form the western defenses of America against the advance of the Pacific.

27. *On the Eastern Side of North America* is the Appalachian System, reaching from the Southern States to the Gulf of St. Lawrence, and giving to the east coast of North America its principal direction north-east and south-west.

28. *On the Eastern Coast of South America* the mountains of Brazil run parallel with the Appalachian System of North America, and secure a parallelism between their corresponding coast lines; namely, that from Newfoundland to Florida Strait, and that between Cape St. Roque and the Strait of Magellan.

29. The *Western Continent is laid out* in two great triangles, North and South America, (*See Illustration on page* 11.) Greenland has a similar shape. This peculiarity is also noticeable in the Eastern Continent, concerning its peninsulas; Africa, Hindoostan', Farther India, Corea, Kamtschatka (*kahm-chat'kah*), Italy, and the Scandinavian peninsula, comprising Norway and Sweden.

30. What is the general direction of the eastern coast of the Eastern Continent? (See Map on page 10.) Of the coast from the south-eastern part of Arabia to the southern cape of Africa? Of the western coast from North Cape to Cape Verd? Of the eastern coast of Hindoostan? From the eastern shore of Greenland to the Gulf of Mexico? From Cape St. Roque to Cape Horn?

What is the general direction of the Pacific coast of the New World from Behring (*be'ring*) Strait to Cape Horn? Of the South American coast from the Caribbean Sea to Cape St. Roque? Of the coast of Africa from Cape Verd to Cape Good Hope? Of the western coasts of Hindoostan' and Farther India?

Mention the principal coast lines which are parallel with each other, and have a north-easterly and south-westerly direction; those which have a north-westerly and south-easterly direction.

31. Hence, it is observed that the *General Directions of Coast Lines* are but two; namely, from north-west to south-east, and from north-east to south-west.

32. Refer to the *Maps* and you will see, furthermore, that such are the directions of nearly all the coast lines of the large islands, peninsulas, and groups of islands in the most important seas, gulfs, bays, lakes, and rivers.

33. *Australia is enclosed* by a coast line composed of six sides, all of which point in one or the other of these two directions.

34. *Above the Parallel of 40° N. Latitude* are the greater parts of North America and Asia, and nearly all Europe; while below the parallel of 40° S. Latitude extends no part of the Eastern Continent, and only the southern extremity of the Western Continent.

35. *Toward the North Pole the Land extends* and expands, as if the Southern Hemisphere was to be surrendered to the ocean; and as new land is being constantly formed in northern latitudes by volcanic action, in time the Northern Ocean may become a land-locked sea.

36. The *Arctic Ocean is connected with the Pacific* by Behring Strait, less than sixty miles in width. Indeed, the Aleutian Isles, which even now reach from Alaska to Kamtschatka, may soon, by means of their fifty active volcanoes, become a continuous rock, joining the two continents, and thus cutting off communication between the Pacific and Arctic Oceans.

37. The *Space between Greenland and Norway*, or between Greenland and Scotland, is no greater than that over which the Aleutian Isles are now being extended. It has already its stepping-stones of Iceland, the Faroe, Shetland, Orkney, and other isles, all of which have been raised by submarine forces yet in operation.

38. The *Longest Straight Line* that can be drawn on the land-surface of the earth would extend north-eastward from Cape Verd to Behring Strait, a distance of about 11,000 miles.

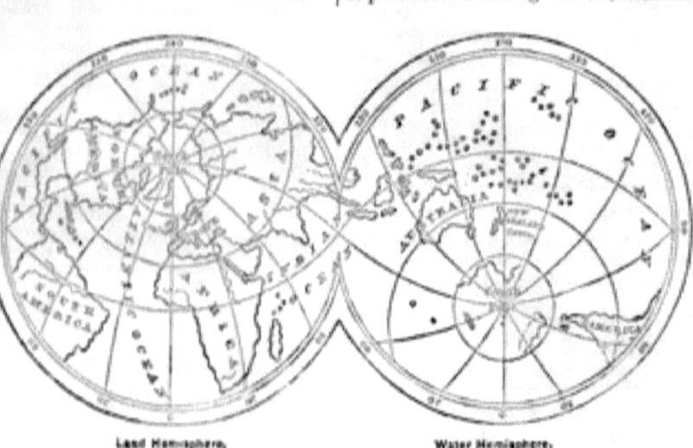
Land Hemisphere. Water Hemisphere.

39. What division of the earth is in the center of the Land Hemisphere? What two divisions are wholly in that hemisphere? What division is almost entirely in it? What part of Asia is in the Water Hemisphere? What division extends farthest into the Water Hemisphere? In which hemisphere is the greater part of South America?

What islands in the center of the Water Hemisphere? Name the largest bodies of land in that hemisphere. In which of these hemispheres is the greater part of the Pacific Ocean,—the Atlantic,—the Indian?

Chart showing the Correspondence between the West Coast Line of the Old World and the East Coast Line of the New World.

Imagine the Old World to be moved westward till the mainland would meet that of the New World: what African gulf would be entered by the eastern part of South America? What American sea by the western part of Africa? Where would be the points of contact? Into what would the Amazon River flow? With what American peninsula would the British Isles be merged? Great Britain would be in what direction from Newfoundland?

40. *An Important Point of Difference* between the divisions of the continents consists in the comparative length of coast lines. In proportion to the extent of surface, the longest line of coast belongs to Europe, the next to North America, and the least to Africa. Europe, with but three sides bounded by water, has, proportionately, four times as much coast line as the whole of Africa; North America has three times as much as Africa.

41. *About One-third of the Entire Land of Europe* consists of peninsulas and islands; and, through the medium of numerous arms of the sea, this division *receives* and *bestows* strength, power, and prosperity; while the closed doors of the African coast forbid entrance to vast regions yet unexplored.

42. *To its remarkably Irregular Coast Line*, together with its mild climate and position on the globe, does Europe owe its greatness among the divisions of the earth.

43. Except in the north, *Africa has no such important Inlets* from the ocean, as those of Europe, North America, and Asia.

44. *Seas, Gulfs, Bays, and Lakes are most numerous* within a belt around the earth, embraced between the parallels of 30° and 60° north latitude.

45. *This Belt*, which is midway between the Equator and the North Pole, *comprises* the most enlightened, powerful, and progressive nations of both continents; here the human race had its origin, here is the birth-place of Christianity, and here flourished nations renowned in ancient history, which were those of Western Asia, Southern Europe, and Northern Africa. Therefore, the superiority of the land divisions of this section is owing, mainly, to the *influences* of their form, position within the North Temperate Zone, and the distribution of their inlets.

46. *Within this Belt*, the inlets on the coasts of the United States, British America, Western and Southwestern Europe, are numerous and important.

Mention the principal bays, gulfs, and sounds on the Atlantic coast of the United States.

Mention the principal seas in Western and Southwestern Europe. Mention the principal bays, gulfs and channels.

47. The *Condition of a Race or People* is affected by contact with surrounding nations and influences; and the greater the facilities for communication and inter-communication, the greater is the advancement; hence, inlets, rivers, canals, and railroads promote the civilization and progress of man.

A City.—River.—Harbor.—Railroad.—Commerce.—Agriculture.

48. *Asia and Europe together form* a vast peninsula, which, with that of Africa, composes the Eastern Continent.

49. *Were it not for a Separation of Sixty Miles* between the Mediterranean and Red Seas, each of these peninsulas would be a vast island or continent.

In this respect, what similarity exists between the Old and the New World? Were the isthmuses of Darien and Suez overflowed, how many and what continents would there be?

50. The *Peninsula comprising Europe and Asia* has its greatest extent from Behring Strait on the north-east to Portugal in the south-west, a distance of about 8,500 miles, or one-third the earth's circumference. It is remarkable for the number and extent of its indentations, which give to it the appearance of a great plant, extending its numerous roots in all directions for nourishment and strength.

51. *This is not the case*, however, with South America, and still less with Africa, which is like a plant almost destitute of roots.

Mention the principal indentations of Europe; the peninsulas formed by them; the seas, bays, and gulfs of Asia; the peninsulas.

52. *Europe extends* from the foot of the Ural Mountains westward, over a great expanse of land,—a continuation of the northern plain of Asia,—to the Carpathian Mountains and the Baltic Sea. Beyond these limits it becomes narrow; facilitating external and internal communication.

53. The *Coast Line is so greatly diversified* by the penetrating arms of the Atlantic Ocean and the Mediterranean Sea that nearly all western and south-western Europe is composed of peninsulas.

TABLE SHOWING THE COMPARATIVE EXTENT OF COAST-LINE.

Grand Divisions.	Square Miles.	Length of Coast Line.	Square Miles for 1 of Coast.
Europe	3,830,357	17,000	229
North and Central America	9,050,927	24,000	345
South America	6,054,131	13,600	477
Asia	16,415,738	35,000	500
Africa	11,516,650	16,000	741

54. The *Three great Land Divisions of the South*,—Africa, South America, and Australia,—resemble each other in their lack of sea arms, and in their backwardness of development; presenting, in these respects, a strong contrast to the divisions of the North.

55. The *Western Continent has its greatest Extent* from the northern part of Russian America south-eastward to the Strait of Magellan, a distance of about 10,000 miles.

56. The *Northern and North-eastern Parts of N. America* are remarkable for their great number of inlets from the sea, cutting the land into a great variety of islands and peninsulas.

57. *Baffin Bay separates Greenland* from the main land of the Western Continent, and *Hudson Bay* forms the great peninsula of Labrador and East Main.

58. *As you go South, you meet* the Gulf of St. Lawrence, Gulf of Mexico, Caribbean Sea, and the Mouth of the Amazon.

59. *Characteristic of the Atlantic Coast of the United States*, are its numerous bays and other inlets; the principal being the Chesapeake, Delaware, New York, Narragansett, Massachusetts, and Penobscot Bays; besides Long Island, Pamlico, and Albemarle Sounds. On *the Pacific Coast*, the most important inlets are San Francisco Bay and Puget's Sound.

60. *South America has its entire North-eastern Side turned* toward Europe and North America, as if to invite their aid in its development; and, although joined by land to North America, the water affords far easier communication than the mountainous region of the isthmus.

61. *Had the Wide Pacific rolled between Europe and America*, instead of the narrow Atlantic, Columbus would probably not have discovered America; *or, had the great Mountain System* of America been placed on the eastern coast, shutting out the Atlantic as it now does the Pacific, and presenting to the east the same undeviating coast line that it does to the west, the New World would probably be less adapted to the progress of mankind than Africa or Australia.

62. Between the eastern side of the New World and the western side of the Old, there is a remarkable analogy, not only in the parallelism of the general coast lines, but also in their system of seas, bays, and other inlets from the ocean.

Section VI.

THE CONTINENTS,—THEIR RELIEFS.

1. The *Land of the Continents* is greatly diversified,—low in some parts and high in others; the altitude or absolute elevation of a place being the distance above the level of the ocean.

2. The highest mountains, as compared with the size of the earth, are no larger than grains of sand on a globe ten inches in diameter; they nevertheless exert vast influences upon the conditions of the whole land surface of the earth.

3. *Plains elevated but slightly* above the level of the sea are called lowlands, even though hills may rest upon them; those of higher elevations, enclosing and supporting mountains, are highlands or plateaus.

4. The *Transition from Low to High Land* is varied; being either abrupt, gradual, or terraced.

5. *A Mountain Range or Chain* is a succession of mountains which have similar geological formations. The *Highest Point* in a chain is called the culminating point.

6. *A Mountain System* is two or more parallel ranges, connected with each other, or which rest upon the same plateau.

7. The *Soil of the Valleys* is fertile, and the climate generally delightful.

A Valley in Switzerland.

8. Although *Mountains and Plateaus* are both elevations of land, and are connected, yet they should be considered distinct from each other. The rugged, broken outline of lofty mountain peaks, with their intervening valleys and passes, presents a strong contrast to the comparatively dull and even surface of a plateau; just as a deeply indented coast does to one whose line is almost unbroken.

9. *No Precise Height* has ever been prescribed, according to which elevations of land should or should not be called mountains.

10. The *Loftiest Peaks on the Globe* are among the Himalayas, the principal one, Mt. Everest, being over 29,000 feet high. Mt. Aconcagua, the highest in S. America, is 23,906.

The highest peaks of the Rocky Mts. are between 13,000 and 15,000 feet high. The White Mts. are about 6,000, the Catskills 3,000, and the Alleghanies from 1,000 to 5,000 feet.

11. ***A Plateau*** is an extent of land elevated above the level of the sea from 2,000 to 14,000 feet.

12. The ***Surface*** may be level, rolling, or hilly; some plateaus contain mountains, valleys, and lakes.

13. ***Plateaus owe their Elevation*** to internal forces, exerted, not as in the more sudden and violent formation of mountains, but slowly and gradually; giving them a comparatively level and unbroken surface. Should, however, the force from beneath be so violent as to cause ***Openings or Seams*** in the earth's crust (see Illustration, page 8), there would be projected through this fissure melted mineral matter, called lava, besides stones, cinders, and ashes; which, falling and hardening upon the uplifted surface, would form a conical pile called a mountain.

14. The ***Upheaval of Hills and Mountains from the Bottom of the Sea*** accounts for the finding of sea-shells on their sides and tops; and the boulders, stones, pebbles, and gravel found in all countries, were irregular fragments of rock, broken off by violence or by atmospheric action, and carried great distances by the rush of water, ice, and icebergs, from high to low ground.

15. ***Mountains were raised to their Present Elevation*** by violent and repeated convulsions, the process extending over thousands of centuries. It is the opinion of geologists that the upheaval of the highest mountains was more sudden, and attended with more violence than that of the ranges of less elevation; that the Alleghany and Brazilian Mountains were raised more slowly, and in earlier periods, than the Rocky and the Andes Mts. The Alps were upheaved more suddenly, and at a period comparatively recent.

16. ***Mountains which have been violently Elevated are known*** by their deep fissures, and great displacement of strata and fossils.

17. The ***Direction of a Chain of Mountains*** is due to the position of the rent made in the earth's crust.

18. ***Mountain Chains extend*** mostly in either of two general directions; from north-east to south-west, or from north-west to south-east.

19. What chains extend from north-east to south-west? What from north-west to south-east?

20. The ***Pressure from beneath forces up***, also, masses of the earth's crust from a considerable depth. Granite is supposed to form the lower part of the crust; hence its appearance in mountains and other parts of the surface is due to volcanic pressure. Those ***Mountains whose Fissures are not yet filled up*** by the lava from beneath, but continue to emit it, are called volcanoes.

21. The ***Principal Plateaus and Mountain Ranges*** of a continent are between its center and one of its sides, following the general direction of the nearest side, toward which their descent is the most abrupt.

22. The ***Longest and most gradual Descent***, either by a continuous slope, or by successive steps, called terraces, is toward the center of the continent, or the greater mass of land.

23. The ***Rocky***, the ***Andes***, and the ***Scandinavian Mountains*** have their long and gradual slope on the east, and descend abruptly on the west. The ***Himalayas*** and the ***Alps*** descend abruptly toward the south. The highland surface of Spain is terraced from the Pyrenees and Cantabrian on the north to the Strait of Gibraltar on the south.

24. The ***Great Plateau System*** of Asia lies south of the Altai Mountains; that of Europe south of the Baltic Sea; of Africa south of its central part; and of America along the west coast.

25. The ***Climate on Mountains and Plateaus*** is cooler than on the lowlands of the same latitude, and the greater the elevation the lower the temperature: hence, upon the ***Elevation of a Country***, as well as upon its latitude, depend its climate, productions, and to some extent, the pursuits of the inhabitants.

26. ***Elevated Regions serve to moderate*** the temperature of the lowlands adjoining them. When air is heated it becomes lighter than the cooler air above it, and ascends; the cold air descending to take its place.

27. Therefore, as the ***Elevations are greatest in the Hot Regions*** of the earth, and diminish toward the poles, the inhabitants of the sultry tropical plains, at the foot of lofty mountains, are continually refreshed by the cool air which comes down from their snowy summits.

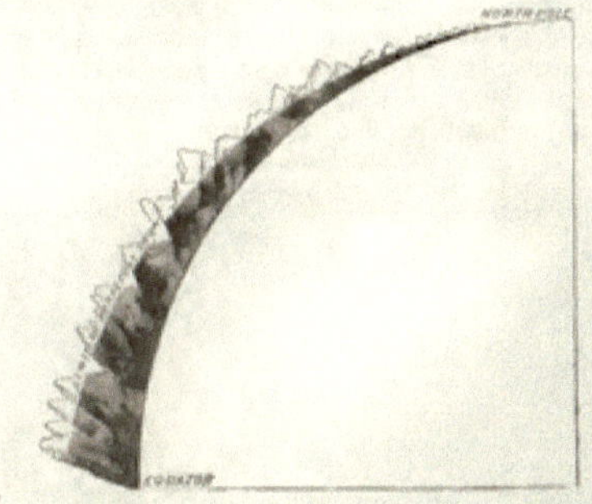

Comparative Height of the Mountains in America, from the Equator to the North Pole; also, the Limit of Perpetual Snow

28. For the same reason that you put a piece of ***Ice into a Pitcher of Water*** in summer, rather than in winter, Providence has uplifted the highest mountains in the tropical, and not in the polar regions of the earth.

29. The ***Most Elevated Plateaus*** on the globe are those in the south of Asia, near the Tropic of Cancer. They have an altitude of more than 15,000 feet above the sea, and on them rest the loftiest mountains in the world; some of the peaks of the Himalayas are more than 28,000 feet above the sea.

30. ***In the Hot Regions of South America*** are the plateaus of the Andes, ranging between 10,000 and 14,000 feet in height, and supporting many peaks between 15,000 and 23,000 feet high.

31. The ***Highest Plateaus of North America*** are in Mexico and Central America, being from 5,000 to 8,000 feet above the sea.

32. *Nearly the whole of Mexico* is a plateau, whose inhabitants, even in the tropical part of the country, enjoy a temperate and healthful climate, owing to its great elevation above the sea.

Section of Mexico from the Pacific Ocean to the Gulf of Mexico

33. The *City of Mexico* is 7,400 feet above the sea level, about twenty times higher than Trinity Church steeple, in the city of New York.

34. *Central and Southern Africa* is one vast table land, the most extensive in the world. It descends on all sides by terraces, to the strip of low ground along the coast.

35. The *Great Mountain System of Europe*, comprising the Cau'casus, Alps, Pyrenees, Cantabrian, and Apennines, is in the southern, or warmest part of that division.

36. *In the Northern Regions of Europe* the only important elevations are the Scandinavian Mountains of Norway and Sweden, which, however, average less than one half the height of the mountains in the south of Europe. With this exception, the northern regions are, comparatively, lowlands.

37. *Take away these lofty Mountain* ranges and extensive plateaus from the places now occupied by them, or remove them from the hot to a cold zone, thus increasing the heat of the tropical and the cold of the frigid regions, and the consequence would be a complete derangement of climates, productions, and the conditions of the inhabitants.

The Alps.—A Glacier.—A Tunnel in the Ice whence issues a Stream which is the commencement of a Large River

38. *In the Tropical Andes, the Region of Perpetual Snow* is above the line of 16,000 feet elevation; *in the Alps*,—Temperate Zone,—it is about 8,500 feet above the sea level; and, *in Arctic Latitudes*, it reaches down to the sea.

39. The *Masses of Snow upon the Mountains* being constantly increased, force their way down the valleys to warmer regions below the snow-line. *By Pressure, alternate Thawing and Freezing* of the upper surface, the whole becomes a great stream of ice, called a *Glacier*, varying in depth from a few hundred to several thousand feet. The *Water that descends through the Crevices* of the ice unites with springs and flows down the mountain sides through tunnels which it cuts in the ice and snow. Every glacier is thus the source of a stream. The *Best known Glacier Region* is that of the Alps.

40. *Draw a Line from the Sea of Mar'mora Northeastward to Behring Strait*, and you will have, south of this line, nearly all the great elevations of Asia, consisting of a vast system of plateaus, supporting lofty mountains whose tops are constantly covered with snow; to the north of these lies the great Siberian Plain.

41. The *Highest Plateau on the Globe* is that of Central Asia, which extends 1,500 miles from the Altai Mountains on the north, to the Himalayas on the south, and 2,500 miles from west to east; having about the same dimensions as the United States, and an average elevation above the sea of 10,000 feet. Its *Surface is greatly diversified* with heights and depressions, rivers and lakes. The principal rivers are the Ganges, Brahmaputra, Indus, Amoor, and Hoang Ho. The Ganges has its two principal sources situated in immense masses of snow, at the elevation of 13,000 feet. The *Elevations diminish* gradually from the Himalayas northward to Siberia, where the slope continues downward to the Arctic Ocean.

42. *Nearly all Western and South-western Asia* consists of plateaus about 4,000 feet high.

43. *This System of Highlands extends* westward to the Atlantic Ocean, over Southern Europe and Northern Africa; the Mediterranean, Caspian, and Black Seas being considered its great depressions.

44. While much *the larger part of Asia consists* of vast plateaus, *Europe consists* mainly of an extended plain, which commences at the Strait of Dover, extends eastward between its great mountain system and the Baltic Sea, and then opens upon and covers Russia. The surface of this plain is almost level, and has an elevation of about 1,000 feet.

45. The *Average Height of the Alps* is between 8,000 and 10,000 feet; the highest peak, *Mt. Blanc*, being over 15,000 feet. The *Apennines* average from 4,000 to 8,000 feet; the *Sierra Nevada* of Spain from 6,000 to 10,000 feet; and the *Scandinavian* Mountains of Norway and Sweden about 4,000 feet.

46. The *Great Plateau of Africa* ranges from 2,000 to 10,000 feet in elevation; its highest part being in Abyssinia.

47. The *Loftiest Peaks in Africa* are Kenia and Kilimandjaro, whose summits are 20,000 feet above the sea.

48. *Central Africa*, north of the Equator, descends to the level of the Great Desert, which is between 1,000 and 2,000 feet above the sea. The highest ranges on the African plateau are the Abyssinian, Cameroon, and Snow Mountains; the highest peaks are Kenia and Kilmandjaro.

49. The *Principal Plateaus of the New World* are in South America, among the Andes.

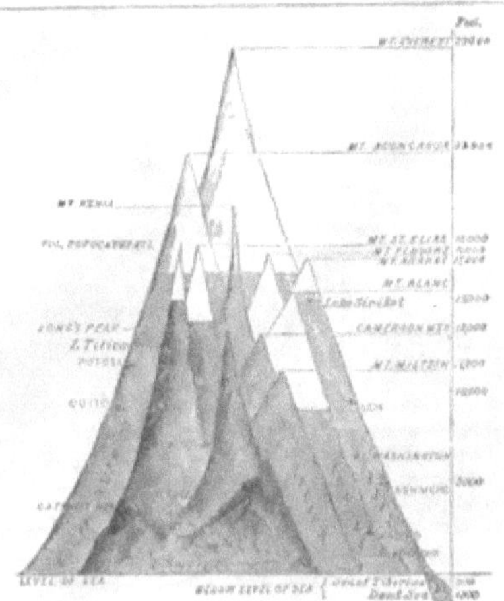

Comparative Elevation of Cities, Mountains, and Lakes.

50. The city of *Quito*, (*ke'to*,) in Ecuador, is built on a plateau nearly 10,000 feet above the Pacific.

51. *Potosi*, a city of Bolivia, is built on a plateau so high that the streets of the city have an elevation of more than 12,000 feet above the level of the sea.

52. *Lake Titicaca*, (*tit-e-kah'kah*,) between Bolivia and Peru, has nearly the same level, being twice the height of Mt. Washington in New Hampshire, four times that of the Catskill Mountains, and seven times that of the Blue Ridge at Harper's Ferry.

53. *High as are these Cities, Lakes, and Table-lands*, yet they are far over-topped by the surrounding mountains, which rise about 10,000 feet above them; hence, these places are but little more than half-way up the highest of the Andes.

54. The *Rocky Mountains, if placed beside the Andes*, would reach only to the plateaus of the latter. The elevation of the *Appalachian* range is only about one-seventh that of the Andes.

55. *On the Western Side of the Andes*, the slope toward the Pacific is abrupt: on the eastern, or Atlantic side, it is gradual; being interfered with only by the Brazilian Mountains, which, however, are less than one-fourth the height of the Andes.

56. The *Andes rise so High* that their tops are in the region of perpetual snow, while, at their foot, the heat is oppressive, and would be greatly intensified, but for their cooling influence.

57. *To the influence of Mountain Systems* do vast regions owe the rain which is necessary to the preservation of vegetable and animal life; for as vapor rises from the earth's surface by the agency of *Heat*, so it must return by the agency of *Cold*. (*See illustration above.*)

What mountain in Asia is the highest on the globe? How far above the level of the sea is Mt. St. Elias, in North America? The City of Potosi, in South America? Great Salt Lake? Mt. Washington? The Catskill Mountains? Madrid, in Spain? How far below the level of the sea is the Dead Sea?

The Andes. Llanos and Pampas. Brazilian Mts.
The Rain of South America.

58. *As the Torrid Regions of the Earth require the greatest amount of Rain*, there are the loftiest mountains, which act as huge condensers of the clouds and vapors floating in the atmosphere; and by the melting of the snow on their sides, they supply springs and rivers to the plains below.

59. *If South America contained no such Elevations*, the quantity of rain poured upon the vast plains would be greatly diminished.

60. In the tropical regions of South America the *Rain-bearing Winds blow*, not from the Pacific, but from the Atlantic Ocean. The clouds, floating westwardly over the land, feel the cooling influence of the Andes, and respond with copious rains, which cover with the heaviest vegetation a region that would otherwise be a sunburnt wilderness.

61. *In Some Districts between the Andes and the Pacific*, rain is almost or wholly unknown, because the clouds are exhausted before passing the mountains.

62. *Had the Andes been raised on the Eastern Side* of that great peninsula, instead of on the western, the rain would fall in torrents upon the then short Atlantic slope, and South America would be deprived of its immense rivers, dense forests, and fertile plains.

63. *Although the Mountain Chains and Plateaus of South America are Extensive*, yet they only cover about one-fifth of its surface, the greater part of it being vast plains.

64. *These huge Piles, called Mountains, projected by Violence* through fearful gaps in the earth's crust, from the melted interior, and occupying such positions of usefulness to the earth and to man, stand in their appointed places, as monuments, not of the Creator's power alone, but also of His wisdom and goodness.

65. *By means of these great Upheavals*, man derives a knowledge of the interior formations of the earth, and obtains the wealth of the mines, which, without these convulsions, would yet lie deep in the earth and beyond his reach. (*See illustration, page 8, second column.*)

66. *Gold, Silver, Iron, Coal*, and other precious and useful minerals were formed below the earth's surface.

Mining.

arkable Features in the formation at deserves notice on account of it is *their Formation in Peaks,* pendicular sides, *Passes* are left. on a plateau, is termed by the semblance to a *saw.*

nd you will have a good illustra- and a plateau; the fingers, sepa- ent the mountain peaks, and the

deprived of their Land Eleva- the climates alone would render or the abode of mankind.

h Mountain Ranges connected, arrier, nations on opposite sides apart in their relations with each between them.

hains there are Natural Passes mountains.

e Alps are not half-way up the onately lower than those of other

tional Communication are now recognized by the Creator when h their intervening passes.

rface of the Earth were made is with the material from the ele- bout 900 feet above the sea level.

ing all the Mountain Systems of the polar regions, they would not diameter equal to the equatorial.

ere are Two great Mountain Appalachian, or Alleghany.

ystem is supported by the North American Plateau, which is el and extends over a great part o the western third of the Unit of British America.

78. *In this System are inclu* Nevada, and Sierra Madre. E Great Basin, or Plateau of Uta

79. The *Rocky Mountain Sy* of Panama, in a north-westerly at about 70° north latitude.

80. *Its Widest Part* is in th all that region between the Pac of Colorado, a distance of over

81. The *Surface of the Plat* Peak to the Missouri River.

82. The *Most Western Rang* the southern extremity of Lowe the Pacific coast as far north a

83. The *Sierra Madre comm* Colorado and extends into Mex

84. The *Most Northern Pass* the Rocky Mountains, is near t and Lewis Rivers, and is one of road to the Pacific.

85. TABLE SHOWING THE CULMI ELEVATION O

	MEAN ELEVATION. Feet.
Asia..................	1,080
South America.......	1,060
North America.......	700
Europe...............	630

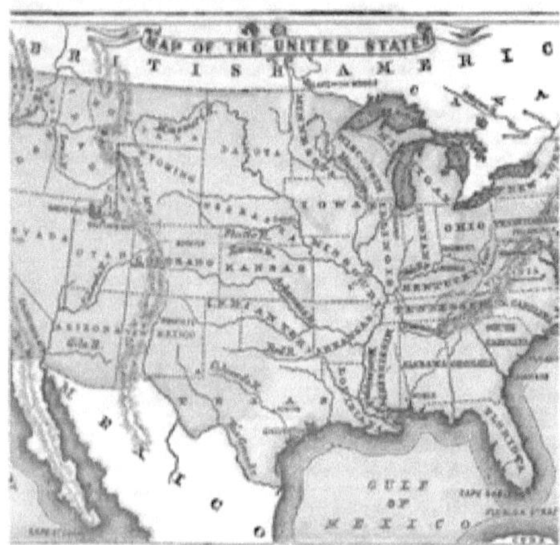

Mountains and Plateaus of the United States are here colored Brown; the Lowlands and Valleys

Section VII.

VOLCANOES,—EARTHQUAKES.

Volcanoes, and Fissures caused by Earthquakes, may be Illustrated by means of a Cake which is burst open at the Top, by the Escape of Steam arising from the Fluids within the Cake; the Heat of the Oven corresponding to that of the Earth's Interior.

1. *Volcanoes, Earthquakes, the Rising and Sinking of the Land* are all attributed to the pressure of steam and gases, proceeding from the heated interior of the earth.

2. *A Volcano* is an opening in the earth's crust through which issue melted rock, or lava, stones, ashes, flame, smoke, and steam. (*See Illustration on page 8.*)

3. The *Materials thrown out usually accumulate* around the opening, called the crater.

4. A *Rent in the Earth's Crust* may be made beneath the sea, where a high mountain will sometimes be formed; sometimes no elevation appears; the fire, lava, and other material being thrown upwards through the water.

5. *Volcanoes allow the Escape of* fire and gases from the interior of the earth, and thus prevent greater destruction by earthquakes.

6. *Some Volcanoes remain inactive* for long or short periods; some now called extinct may again become active.

7. Volcanic action is usually preceded by earthquakes, which sometimes rend the earth open in fissures, and engulf whole villages and cities.

8. *By these Convulsions Mountains and Hills are raised,* in some instances, from what, a few hours before, were low lands.

9. *Through the Openings issues Volcanic Matter.* Vast quantities of water, mud, and fish are sometimes ejected from mountains in South America; they proceed from subterranean lakes or pools.

10. On the *West Coast of Italy* a volcano called Monte Nuovo (*noo-o'vo*), over 400 feet in height, was formed in two days, and the adjoining land was elevated so that the bottom of the bay for 200 yards from the shore, was raised above the water level, leaving the fish on the newly formed shore to be picked up by the inhabitants.

11. The *Most Noted Volcanoes in the World* are Vesuvius on the coast of Italy, Etna in Sicily, Stromboli (*strom'bo-le*) on one of the Lipari (*lip'a-re*) Islands, Hecla in Iceland, Cotopaxi (*ko-to-pax'e*) one of the Andes, Sangay near the city of Quito, Mauna Loa on the island of Hawaii (*hah-wi'e*), and Teneriffe on one of the Canary Islands.

Fissures caused by an Earthquake in Italy 1783.

12. The *Number of Active Volcanoes on the Earth* is about 250, more than half of which are on the coasts and islands that line the Pacific Ocean. The most remarkable volcanic region is in Malaysia. Continents have their volcanoes mostly on their borders; those of the Western Continent are chiefly among the Andes and the Rocky Mountains.

13. Although *Earthquakes mostly occur in Volcanic Districts*, yet any part of the earth's surface is subject to them. Some are violent and destructive, while others are almost or entirely imperceptible.

14. *On the Western Continent, Earthquakes are most frequent* in Central America, Chili, and Peru; in Europe, they occur chiefly in Italy and its vicinity.

15. The *Approach of an Earthquake*, like the eruption of a volcano, is sometimes indicated by symptoms of unusual agitation beneath the surface of the ground.

16. *Among the Greatest Earthquakes* of which we have a record, is that which destroyed the cities of Herculaneum and Pompeii (*pom-pay'e*), A.D. 63; and, after they had lain in ruins for sixteen years, they were again overwhelmed by an eruption of Mt. Vesuvius.

17. In 1692, *Port Royal, the Capital of the Island of Jamaica*, was sunk in less than one minute; the sea rolling in, and driving the vessels that were in the harbor over the tops of the sinking houses.

18. The *Great Earthquake of Lisbon*, in 1755, commenced with a dull, rumbling sound below the surface, immediately followed by a tremendous shock, which threw down a large part of the city; and, in the space of six minutes, 60,000 people perished. The *Sea retired* to a distance, only to return in a vast wave, fifty feet high. The *unfortunate People* rushed from the falling buildings to secure shelter on the new and massive marble quay (*ke*), which suddenly sank with them into the sea; the water closing over the spot to the depth of 600 feet. *Not a single Fragment of the many Vessels*, nor one of the thousands of human bodies that were drawn into this frightful chasm, ever floated to the surface; all being engulfed

in the fissures which opened, and immediately closed over them. A *Portion of the Earth four times as large as Europe* was affected by this terrific shock. The *Waters of the Scotch Lakes* suddenly rose above, and then subsided below, their level. On the *Shores of the West Indies* the tide rose twenty feet, and the water resembled ink; even the coast of Massachusetts and the waters of Lake Ontario were sensibly affected.

19. In 1811, occurred the *Earthquake of New Madrid*, in Missouri, which was remarkable for the continuous quaking and rending, over an extent of 300 miles, during several months. *Great Openings* were made in the surface, from which mud and water were projected.

20. *These Internal Convulsions continued* until they culminated, March, 1812, in the *Earthquake of Caracas*, on the northern coast of South America, by which the whole of that splendid city became instantaneously a mass of ruins, and thousands of its inhabitants perished.

21. In 1822, *an Earthquake occurred in Chili*, which resulted in the elevation of a large section of country to a height varying from two to seven feet.

22. In 1857 and 1858, *Repeated Shocks were felt*, at intervals, in the country around Naples. Several towns were reduced to heaps of ruins, and about 30,000 inhabitants perished.

23. *During the Earthquake*, Mt. Vesuvius continued in action; and, by affording a means of escape for the confined gases, doubtless prevented the entire destruction of the city of Naples and the ruin of all the region in the immediate vicinity of the volcano.

24. In 1859, *the City of Quito* (*ke'to*) and several towns in its vicinity were almost entirely destroyed by an earthquake.

25. In August, 1868, *an Earthquake occurred in Peru, Chili, and Ecuador*, which caused a fearful loss of life and property; and, in October, several shocks were experienced in California, causing considerable damage in the principal cities.

26. *Shocks have been felt* at different times in various parts of the United States.

27. *Subsidences, like Upheavals*, sometimes occur so gently that the inhabitants are only aware of the change by the difference in the sea level. In 1819, an area of 2,000 square miles about the mouth of the Indus, in Hindoostan, was suddenly converted, by an earthquake, into an inland sea.

Fort Sindree before it was Submerged by the Earthquake of 1819.

Fort Sindree after the Earthquake.

28. The *Fort and Village of Sindree sank* so much that only the tops of the fort, houses, and trees were seen above the water.

29. The *Coast of Sweden* has been rising for many years; near Stockholm, at the rate of a few inches in a century.

30. *In Greenland*, the south-west coast has been slowly sinking for four centuries past.

31. In 1866, *an Island was Upheaved* from the bottom of the sea south-east of Greece. The water was violently agitated, and from the fissures rushed flame, smoke, lava, and fragments of rock.

Section VIII.

PLAINS AND VALLEYS.

1. The *Land Surface of the Earth* may be divided into two general classes, highlands and lowlands; the highlands comprising mountains and plateaus; the lowlands, plains and valleys.

2. *Lowlands comprise* all lands whose elevation is not more than 1,000 to 1,500 feet above the sea.

3. *A Plain surrounded by Mountains* or hills is called a valley.

4. *Through the Lowest Part of a Valley*, or near its center, generally flows a river, which drains it.

5. *Lowlands comprise* far the greater part of the land surface of the earth, and in them is found the great mass of vegetation, animals, and mankind.

6. The *Soil of the Lowlands* is constantly enriched by the alluvial washings from the mountain sides, which have filled the fissures and depressions of the rock that originally formed the land surface of the earth; it is still further enriched by the collection upon it of decomposed vegetable and animal substances. About two-thirds of the Western Continent are covered by plains.

7. The *Great Central Plain of North America* is all that part north of the Gulf of Mexico and between the Rocky and Alleghany Mountains, an area of about three and a quarter millions of square miles; comprising four great basins, drained

by the Mississippi, St. Lawrence, and Mackenzie Rivers, and Hudson Bay.

8. The ***Lowlands of South America*** comprise those of the Orinoco, Amazon, and La Plata Rivers, and cover four-fifths of the surface east of the Andes.

9. The ***Lowlands of the Orinoco, termed Llanos,*** are less than 300 feet above the sea level, and present a surface almost as even as that of water. ***During the Dry Season,*** from May to November, the ground is parched and barren; presenting the appearance of a desert. ***During the Wet Season,*** from November to May, the clouds, driven westward by the Trade Winds, pour down their rain; when horses, cattle, serpents, and alligators suddenly appear in vast numbers.

10. The ***Plains or Lowlands of the Amazon, termed Silvas,*** extend from the Andes to the Atlantic, a distance of 1,500 miles, and average 600 miles in breadth. ***They cover an Area*** of about 2½ millions of square miles, and consist, chiefly, of dense forests into which man has scarcely penetrated.

11. The ***Plains of the Amazon*** are about two-thirds the size of all Europe.

12. The ***Valley of the La Plata*** consists mainly of vast grassy flats, called ***Pampas,*** where vast herds of cattle feed; ***these Animals are hunted for*** their hides, horns, and tallow, which constitute the chief export of that region.

13. The ***Three Plains of South America cover*** an area of 5,000,000 square miles, while all Europe contains but 3,500,000 square miles.

14. The ***Great Northern Plain of the Old World*** lies north of its chief mountain system. ***It Extends*** from the shores of the North Sea and English Channel, eastward, over France, Belgium, Holland, Denmark, Northern Germany, Russia in Europe, Russia in Asia, and Independent Tartary, to Behring Strait; interrupted only by the Ural Chain, which forms a natural boundary between Europe and Asia.

15. The ***Portions of this Great Plain*** which are drained by the tributaries of the North, Baltic, and Black Seas, are famous for their fertility.

16. ***That Part of the Plain bordering on the North, Baltic, and White Seas,*** evidently emerged from the ocean at a much later period than some other parts of the continent; indeed, its elevation is yet incomplete; for many parts of Holland are still below the sea level, and are protected from inundation by means of dikes constructed by the inhabitants.

17. In the ***Region of the Caspian and Aral Seas,*** the surface is also much depressed; some parts being below the level of the sea.

18. ***Until a Period comparatively Recent,*** it probably formed the bed of a great inlet, or arm of the ocean, from which it has been isolated by the upheaval of the surrounding highlands. The ***Soil*** contains sand, sea-shells, and salt, and the region is consequently desolate. There being no outlet to the enclosed waters, the seas of this basin are strongly impregnated with salt.

19. ***Toward the Arctic Ocean,*** the plains in Europe and Asia are a boundless waste, swampy in summer, and frozen in winter.

20. The ***Polar Regions of North America*** may be considered a continuation of the lowlands of Northern Asia.

Section IX.

DESERTS AND OASES.

1. ***Deserts are Extensive Tracts*** destitute of water, and, consequently, of vegetation and animal life.

2. ***Their Condition is Attributable,*** chiefly, to the heat and dryness of the winds which blow over them.

3. The ***Desert Region of the Old World*** extends over the greater part of Northern Africa, and north-eastward over vast regions of Arabia, Turkey, Persia, Afghanistan, Beloochistan, Independent Tartary, and the Chinese Empire; this is, also, the great rainless region of the world: its area is more than twice that of the United States.

4. The ***Surface of that part of Sahara*** which lies north of Timbuctoo (*see Map of Africa*), thence toward the Atlantic, is a vast sandy waste covered with a coating of salt and sea-shells.

5. At times, the ***Desert is Visited by the dreaded Simoom***—a hot, suffocating wind which drives the burning sand in great clouds furiously over the surface, for great distances.

6. ***To avoid Suffocation,*** travelers throw themselves on the ground with their faces to the earth, stopping their ears and noses with their handkerchiefs until the storm has passed; their camels lie close to the ground and bury their noses in the sand.

7. ***By means of the Winds which Blow over the Desert,*** some houses, villages, and towns have been completely covered with the driven sand. ***There have been Discovered*** remains of ancient temples so long buried that no record of them is found in history.

8. ***Large Portions of the Great Desert*** are diversified by hills and mountains, between which are valleys or immense tracts either of sand or naked rock.

9. ***Between Fezzan and the Southern Side of the Desert,*** some tribes live on the mountains, at elevations where the temperature requires them to wear warm clothing, even furs. Here, also, rain occasionally falls; while in other districts, the mercury in the thermometer rises to 132° in the shade and 156° in the sun.

10. ***Sahara is a vast Plateau*** which has an elevation above the sea of 1,200 to 1,500 feet. It is about 1,000 miles wide and 3,000 miles long; covering an area equal to about four-fifths that of the United States.

11. The ***Oases*** are fertile spots in various parts of the desert, ***where are found*** springs of cool and delicious water, besides grass, the palm, fern, acacia, and other trees; here travelers and their camels find shade, refreshment, and rest.

12. The ***Oases are Depressions*** in the table land of the desert; the water is supplied from the surrounding cliffs, and is retained by a stratum of clay in the center of the valley.

13. The ***Number of Oases in Sahara*** is about thirty; of which, twenty are inhabited.

14. The ***Principal Desert in the New World*** is that of Atacama, where rain has never been known to fall. It is situated in Peru and Bolivia, west of the Andes. Its dry surface of sand and rock supports not the slightest vegetation.

The Ocean.—A Storm. Some of the Uses of Water. The Ocean.—Fair Weather.

Section X.

THE OCEAN: ITS EXTENT AND DIVISIONS.

1. The *Existence on the Earth's Surface* of a vast body of water is essential to life; for, in the composition of both vegetable and animal bodies, the chief element is water.

2. *Water forms* more than five-sixths of the animal body, and nearly the whole of the vegetable.

3. All *Lakes, Streams, Springs, Rain, and Clouds,* besides all vegetables and animals are, consequently, dependent upon the ocean, which is the great reservoir whence all the land on the earth's surface receives its supply of water.

4. *Influenced by a certain degree of Cold,* water becomes ice; and, influenced by heat, it takes the form of steam and vapor.

5. *Water exists* not only on the earth's surface, but also in the air above the surface, and in the ground below it, where it forms subterranean lakes and streams.

6. The *Water of the Ocean is preserved Pure* by its saltness and constant motion. Fresh water is that which has been raised from the ocean by evaporation, and returned to the land by condensation.

7. The *Sea or Ocean has Five Divisions,* called the Pacific, Atlantic, Indian, Northern, and Southern Oceans.

8. *It affords an Easy Communication* between nations, for their mutual development and prosperity.

9. As there are *Two Great Bodies of Land,* the Eastern and Western Continents, so there are two principal oceans corresponding to them, in both size and shape; the Pacific to the Eastern, and the Atlantic to the Western Continent. The Indian Ocean may be considered a part of the Pacific.

10. *In America, the Mountain Ranges correspond in Size* to the oceans nearest them; the Andes and Rocky to the Pacific, the Appalachian and Brazilian chains to the Atlantic. The highest peaks of the Andes border on the widest pa[rt of] the Pacific.

11. The *Largest Ocean* is the Pacific, which contains al[most] one-half the water on the globe, and covers one-third of [the] earth's surface. It extends from Behring Strait to the [Ant]arctic Ocean; its western shore being Asiatic, and its east[ern] American.

12. The *Shape of the Pacific and Indian Oceans* is [the] reverse of that of the continents, being narrow in the n[orth] and wide in the south.

13 AREAS OF THE OCEANS.

	Square Miles.
Pacific	66,000,000
Atlantic	35,000,000
Indian	30,000,000
Arctic	6,000,000
Antarctic	4,500,000
Total	141,500,000

14. While the Pacific is distinguished for its size, the At[lan]tic is distinguished for its numerous arms which penetrate into the land of both continents.

15. *Owing to these Arms, and the Position* of the Atla[ntic] between the important sides of the continents, this ocean [con]tributes far more than any other to the interests of mankin[d].

16. Mention the principal arms of the Atlantic on its eastern side; o[n its] western.

Into which of the grand divisions do they mostly penetrate?

In what zone are most of these arms?

Has the Pacific such arms on both sides? On which side are its prin[cipal] arms? Mention them. Mention those of the Indian Ocean.

17. The bed of the sea, like the surface of the continent[s, is] diversified by highlands and lowlands; the submarine plat[eaus] causing shallow water, termed shoals and banks.

18. *Near some Coasts,* the ocean is shallow, its bed b[eing] the submerged border of the continent; but, at a dist[ance] from the coast of 100 to 300 miles, the water becomes [sud]denly deep. (*See Illustration on following page.*)

A Sectional View from the Atlantic Coast of the United States eastward and northward, showing the bed of the Ocean, the Comparative shallowness of the Water near the Coasts, the Depth of the Ocean, as compared with that of the Irish and North Seas,—a Steamer sailing from Europe to the United States,—Fishing Vessels off the Coast,—a Wreck at the Bottom of the Ocean,—Whales,—Seaweeds.

19. The ***Depth of the Water surrounding the British Islands*** and the islands east of Asia, is only about one-fortieth of that of the ocean basin.

20. If the ocean were withdrawn from the earth, its bed would appear chiefly as extensive valleys of various depths, and the parts adjoining the continents, as plateaus, sloping suddenly downward to the valleys.

21. The ***Ocean is Deepest*** near the tropics; here, also, are the highest mountains.

22. The ***Depth of the Ocean*** varies from 1,000 to 30,000 feet. Between Ireland and Newfoundland the bed of the sea is a submarine plateau, remarkable for its comparative evenness, and the quietness of the waters that rest upon it. The depth of the water there varies from 10,000 to 15,000 feet.

23. The ***Depth of the Gulf of Mexico*** is about 5,000 feet in its deepest part; of the Mediterranean from 3,000 to 9,000 feet; of the North Sea, 180 feet. The mean depth of the Ocean is estimated to be between 15,000 and 20,000 feet.

24. ***A Depression of the Water Level*** of about 300 feet would extend the main land of Europe and Asia over their neighboring seas and islands.

25. Were the ***Mass of Water diminished,*** so that its greatest depth would not exceed 5,000 feet, the elevation of the continents would be so increased that the climate of the lowlands, even in the temperate and torrid zones, would cause them to become frozen wastes; the most fertile plains of Europe would then have an elevation above the depressed ocean level of over 15,000 feet, the present height of Mt. Blanc; the Mississippi valley would attain a far greater elevation than the present altitude of the highest peaks of the Rocky Mountains.

26. Therefore, it is plain that ***the Climate of any Locality depends*** essentially, not only upon its distance north or south of the Equator, but also upon its elevation above the level of the sea.

27. The ***Saltness of the Ocean*** is supposed by some to be caused by great masses of salt, forming parts of its bed, or by the salt brought into it by rivers; others hold that it was originally made salt by the Creator.

Section XI.

MOVEMENTS OF THE OCEAN.

1. The ***Movements of the Oceanic Waters*** are of three kinds,—waves, currents, and tides. Waves may be influenced by tides or by winds. The tide affects the whole depth of the ocean; the wind affects the water nearer the surface.

2. ***Currents and Tides*** are regular and constant.

3. ***Tides are caused*** by the influence of the moon and sun; mostly of the former.

4. The ***Oceanic Currents are caused,*** or modified, by the winds, the difference of temperature between the Equator and the poles, and by the revolution of the Earth on its axis.

5. ***If the Earth were at Rest,*** the whole surface covered evenly with water, and under no external influence, there would be no currents, or important movements of the water; but admit the warm rays of the sun, and there would follow two great movements; the warm tropical waters flowing toward the poles, and the waters of the polar regions toward the Equator.

6. ***As Cold Water is Heavier than Warm Water,*** the latter would leave the Equator as surface or upper currents, and the cold water would approach it as under currents. Under these circumstances, the directions of the currents would be north and south. Besides this, ***the Water which is taken up from the Tropical Regions by Evaporation,*** is replaced by water flowing from the direction of the poles.

7. Allowing the ***Earth to Revolve on its Axis*** from west to east, and, remembering that the motion of the

surface is most rapid at the Equator and diminishes toward the poles, you will observe that as the waters from the polar regions approach the Equator, they are unable to acquire the more rapid motion of that part of the earth; consequently, the *Water falls behind*, and presents the appearance of a current rushing from east to west, round and round on each side of the Equator; this is called the *Equatorial Current.*

8. The *Course of the Equatorial Current is changed* by the deep sea-slopes of the continents and islands. The eastern angle of South America is so situated that the Equatorial Current is divided at Cape St. Roque.

9. The *Northern Section* of the Equatorial Current here takes a north-westerly direction, enters the Gulf of Mexico between Cuba and Yucatan, and issues from it between Cuba and Florida, and then turns north-eastward, constituting the Gulf Stream.

10. While the *Equatorial Current* appears to seek a westerly direction, it actually moves with the earth eastward; and, although not fast enough to keep up with the unyielding *land* of the Equatorial regions, still, when transferred to those parts of the surface whose easterly motion is less rapid, the Equatorial Current retains sufficient of its *actual* easterly velocity imparted to it when near the Equator, to go ahead of those parts nearer the poles.

Boat Race Illustrating Currents of the Ocean. The Starting Points are shown by the Three Outline Figures on the Left.

11. *When you are on a Steamboat*, its motion causes the water, rocks, and trees near by to appear as if rushing past you in the opposite direction; even when you pass a boat which is sailing in the same direction with you, but *less rapidly*, it appears to move behind and away from you.

12. *In the Illustration above, the Steamboat represents* the land of the Equatorial regions; the small boat in which are two oarsmen, represents the water of those regions. Although both started together as shown in the left of the picture and moved in the same direction,—from west to east,—the swifter motion of the steamboat causes it to leave the oarsmen behind; consequently they appear to the people on the steamboat to move in the opposite direction,—from east to west.

13. The *Two Oarsmen represent* the Equatorial Current; they *actually* move *eastward*, but *apparently westward.*

14. Now compare the motion of the boat containing the two oarsmen with that of the boat containing but one, and it will readily be seen that the former goes ahead of the latter, and moves to the *east*; here, the two oarsmen represent the *Return Equatorial Current* flowing eastward, which in the North Atlantic is called the *Gulf Stream*, while the one oarsman represents the regions toward the poles, where the eastward motion of the Earth on its axis is slower than at the Equator.

15. The *Waters of the Equatorial Current* and the Gulf Stream are warmer than the other waters of the ocean, and have an important bearing upon the climate, productions, and inhabitants of the countries coming under their influence.

16. To the Gulf Stream *Europe is greatly Indebted* for its healthful climate, rich productions, and the general prosperity of its people.

17. The *Numerous Inlets from the Sea* which give to Western and Southern Europe an exceedingly extensive coast line, are peculiarly fitted for the distribution of the favorable influences of the Gulf Stream.

18. *Disconnect North and South America* by an extension westerly of the Caribbean Sea or the Gulf of Mexico, so that the Gulf Stream would flow into the Pacific, and the prosperity of Europe would be suddenly diminished; the *Mild and Genial Climate of the British Isles and France* would be exchanged for that of the bleak coasts of Labrador and Newfoundland, which lie between the same parallels.

19. In the same manner, the *Equatorial Current of the Pacific* continues westward until it reaches the islands east of Asia, where the northern part of the current is turned north-eastward to higher latitudes, where its easterly velocity predominates.

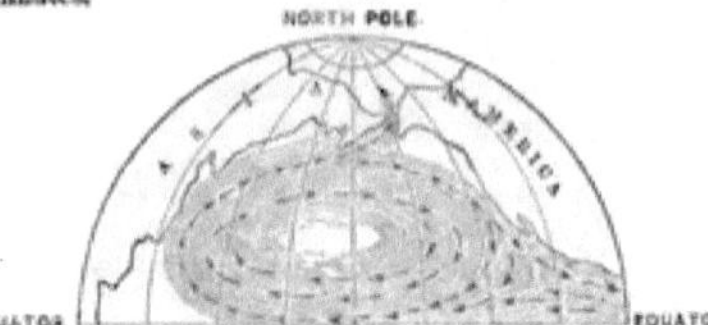

The Equatorial and Japan Currents of the Pacific Ocean.

20. *Under the Name of the Japan Current* it then flows eastward across the Pacific, until turned by the western side of North America, when, following the direction of the coast, it meets the Equatorial Tropical Current.

21. Therefore, the *General Plan of the Equatorial Current* is a flow round and round in ellipses, westward on or near the Equator; turning to the north in the Northern Hemisphere, and to the south in the Southern Hemisphere.

22. The *Equatorial Current flows in Deep Waters*, and its course is bent by the steep sides of the ocean's bed, about 100 miles from the coast line.

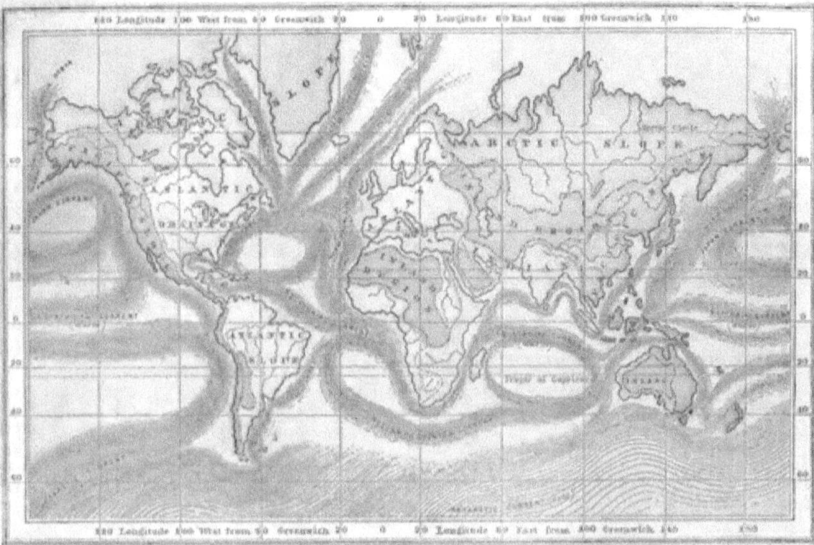

23. *From the Arctic to the Atlantic Ocean* two cold currents flow southwardly; one being west, the other, east of Greenland. These are called Arctic Currents; and, being unable to acquire the easterly velocity of those parts of the earth's surface which they pass on their way south, they are thrown to the west side of the ocean.

24. The *Arctic Currents carry* with them huge icebergs; many of which, as they meet the warm waters of the Gulf Stream off the coast of Newfoundland, become melted, and there deposit quantities of gravel, sand, and stones, transported from more northern lands.

25. *These Masses contribute* to the formation of the famous banks or shoals of that region.

26. Here, also, the *Cold Currents of the Atmosphere from the North* meet the warm, moist air over the Gulf Stream, whose vapors thus become condensed and form the heavy fogs for which that region is noted.

27. *By means of these Currents*, there is maintained a constant interchange of tropical and polar waters; thus moderating the heat of the Torrid Zones, and the cold of the Frigid.

28. The *Difference of Temperature* between the waters of the Gulf Stream and those which wash the east coast of North America, is, in winter, between twenty and thirty degrees; and, the climate on the eastern coasts of the Atlantic, at the latitude of 60° is as warm as that on the west coast, at the latitude of 40°.

29. *Even in Winter, the Gulf Stream carries the Temperature of summer* as far north as the Banks of Newfoundland. Evaporation from its warm waters is very rapid, hence the dampness in the atmosphere of the Atlantic States when easterly winds prevail.

30. The *Gulf Stream*, on reaching the British Islands, is divided; one part entering the Arctic Ocean, while the other is turned southward along the south-western coasts of Europe, where *its effect upon Atmosphere and Climate* is visible in the fertile vineyards and beautiful landscapes of that section.

31. The *Average Velocity of the Gulf Stream* is one and a half miles an hour; off the coast of Florida it is most rapid, being from three to five miles an hour. In the Pacific Ocean the Equatorial Current moves at the rate of about three miles, and, in the Indian Ocean, of two and a quarter miles an hour.

32. *Within the circuit of the Gulf Stream* are large collections of floating sea-weed, giving to the middle part of the North-Atlantic the name of the Sargasso Sea, (*Sargaso*, Spanish for *Sea-weed.*) These collections are caused by the whirling motion of the Gulf Stream.

33. *The Gulf Stream may be traced* throughout its course by the warmth of its waters; and its deep blue color contrasts strongly with the green waters of the Atlantic Ocean.

34. The *Cold Current from the Antarctic Ocean* is divided in its north-easterly course by the south-west coast of South America. One part flows northward into the Equatorial Current; while the other part flows around Cape Horn and takes an easterly direction, toward Australia.

35. The *Equatorial Current of the Indian Ocean* connects with that of the Atlantic by a westerly current which doubles Cape Good Hope, called the Cape Current, in which vessels sail that are bound westward. South of the Cape Current is the return or counter current, in which vessels sail that are bound eastward.

36. *Vessels Navigating the Pacific*, between North America and Asia, sail westward in the Equatorial Current, and eastward in the return flow, called the Japan Current.

37. These two currents together form a great ellipse; its southern side being the Equatorial Current, and its northern side, the Japan Current.

38. *From the Japan Current, a Stream of Warm Water flows Northward* through Behring Strait; this, with a similar current from the Gulf Stream, tends to moderate the cold of the Arctic region, and to balance the cold currents flowing south on both sides of Greenland.

39. The *Climate of a Country depends chiefly* upon its latitude and elevation. It is also affected by the ocean and its currents.

40. The *General Flow of the Ocean Currents*,—westward in the tropical, and eastward in the temperate regions,—coincides with the atmospheric movements. In the tropics the winds blow to the west, and are called Trade Winds; in the Temperate Zones they blow to the east, and are called Return Trade Winds.

41. The *Temperature of the Atmosphere* is regulated by winds, or currents of air; while that of the ocean is regulated by currents of water.

42. Besides the *Great Benefits of the Ocean* already mentioned, there is another, in its myriads of fishes, which afford food and luxury to man; and, it is an interesting fact that the best fish are found in the cold currents, near the coasts.

43. The *Observing Learner cannot fail to see* that the ocean, which to the thoughtless appears as a great waste, is vast in its benefits; for it provides man with rain and streams to bring forth grass, fruit, and grain; tempers climates; bears his ships from nation to nation, and furnishes its living creatures as food for his table.

44. *Were the Warm Currents not turned toward the Poles*, the polar waters, now open, would be continually covered with vast fields of ice; hence, the coasts of America, extending far north and south, and turning the currents in their various directions, were thus formed according to a wise design.

45. The *Unceasing Activity of the Waters* of the ocean contributes largely to the benefit of all vegetable and animal life, and also to their own purity. This law of reciprocity applies with equal force, to nations and to man.

46. *Imagine the Tropical and Frigid Regions to be in a State of Rest*, refusing to exchange their waters; one would be intolerable from excessive heat, the other, from excessive cold; the result would be ruin to both. So, also, would it be with man in a state of idleness.

47. The learner cannot fail to recognize in the Creator's plan for the development of the earth, and for the welfare of its inhabitants, the benefits of the *Great Law of Contrasts*; whereby exist heat and cold, land and water, highlands and lowlands, the mineral, vegetable, and animal kingdoms.

Section of a Hill, whence issues a Spring.
A. Loose Earth or Broken Rock through which the Rain sinks.
C. Solid Rock or Hard Clay not penetrated by Water.
B. Seam or Stratum in which the Water flows.

Section XII.

EVAPORATION, SPRINGS, AND WELLS.

1. *To the Ocean*, although salt, do we owe all the fresh water of the land. It is the source whence all springs, rivers, and lakes are supplied. The ocean and its streams of fresh water throughout the land, resemble the heart and veins by which the life of an animal body is sustained.

2. The system by which the *Land receives from the Salt Ocean a Bounteous Supply of Fresh Water*, is remarkable, as much for its completeness, as for the benefits which it imparts.

3. *All is the effect of* the combined action of heat, cold, and air. Heat lightens the water, that the air may lift it from the ocean; the winds carry it in the form of vapor over the land; the cold makes the vapor heavier than the air, and then it falls in the form of rain, snow, hail, and dew.

4. The *Rain that falls upon the Ground* serves to water the fields, and to fill lakes, rivers, ponds, and cisterns, for man's use. A part of it sinks into the ground, and forms subterranean streams or reservoirs; other portions are evaporated, and they again return, either to the land or to the ocean.

5. *Without Evaporation*, there would be no rain or dew, trees or grass; the whole land surface of the earth would be parched and barren.

6. The *Water which forms our Springs* and fills our wells, is rain which has fallen on neighboring lands, at or above the level of the springs.

7. *Rain-water* percolates through the gravel, loose soil, or fissures in rocks, until stopped by a bed of rock or clay, impervious to water.

8. As the *Upper Side of the Bed* is inclined toward low ground, the water flows in that direction; and finding an opening, it issues forth as a spring.

9. *Whatever Cavities exist* in the upper surface of this bed, become natural cisterns, which preserve the water pure and cool for our use in dry seasons.

Section of the Ground or Rock, showing how Wells are supplied.
A. The Part through which the Rain Water percolates.
C, Rock or Clay impervious to Water.
B, Seam or Stratum in which the Water passes.

10. *Wells are supplied* with water from the stratum in which it rests or flows, or with that which finds its way into them, through the crevices of the rock.

11. *Springs may be supplied by* rain or snow that falls on elevated ground several miles distant.

12. *After a Dry Season,* the flow from most springs becomes diminished, and sometimes ceases, until replenished by rain. There are, however, some springs whose discharge is uniform throughout the year; these are supplied from subterranean reservoirs, too extensive to be materially affected by ordinary droughts.

13. The *Quality of Spring Water* depends upon the materials composing the rocks or soil through which it flows. That which issues from sand-stone rock is softer and purer than that flowing through lime-stone strata.

14. *Intermittent Springs* are those which flow, and cease to flow, during alternate periods throughout the year.

15. *Mineral Waters* are those which possess medicinal qualities, owing to certain mineral substances which they hold in solution. There are, also, springs of salt water.

16. *Mineral Waters* are used for purposes of drinking and bathing. Mineral springs are numerous in the United States; *the most celebrated* are those of Saratoga and Virginia. They abound, also in England, France, and Germany.

17. The *Strata at the Sides of the Continents* being inclined to the ocean, many subterranean streams empty into it, through its bed. In some instances, these streams are forced upward to the surface of the ocean; this is caused by the pressure of the water within the surrounding high grounds. Off the south coast of Cuba, springs of this nature burst upward through the salt water with great violence.

18. The *Waters of Thermal, or Hot Springs,* are those which have penetrated to such a depth as to come in contact with the heated rocks, or lava beds, in the interior of the earth; here, steam is produced which forces hot water and vapor through crevices in the rock, from subterranean pools, up to the surface. The waters of thermal springs are used for the purpose of bathing.

19. The *Most Noted Hot Springs* are those of Iceland, Central France, Asia Minor, Virginia, California, and Yellowstone Park.

Boiling Springs, illustrated by means of a Tea-kettle.

20. *Boiling or Hot Springs* may be illustrated by a kettle partly filled with water, and placed upon a hot stove; the kettle representing the subterranean cavern, and the stove, the heated rocks of the earth's interior. The steam, if prevented from escaping at the top, presses upon the hot water below it, and forces it out through the spout, as shown in the illustration above. When the water in these caverns is long boiled and exposed to great heat, steam may be so suddenly generated as to produce explosion; this may account for the geysers (*ghi'zerz*), or fountains of boiling water.

21. *Geysers* are of various dimensions; some are constantly boiling, others boil up only at intervals, with loud explosions.

22. The *Most Celebrated Geyser Regions* are in Iceland, California, and near the headwaters of the Yellowstone and Madison Rivers in the United States. The geyser region of the Yellowstone and Madison Rivers is more wonderful than any other that has yet been discovered.

The Geysers, Iceland.

23. *Subterranean Streams* produce excavations and subsidence of the soil. Flowing down a hill or mountain, just beneath the soil in which trees have their roots, they sometimes cause considerable tracts of land to slide down from the mountains; these tracts are called landslides.

24. *If the Subterranean Bed of Rock or Clay,* over which the water passes, were at the surface of the ground, instead of some distance below it, the land would be inundated by every shower; or, if so deep as to be far below the surface, springs would not exist, or would be beyond man's reach; and, without springs, rivers would not be kept supplied.

Artesian Wells:—A. A. A. Strata impervious to Water;—B. B. Seams or Strata in which Subterranean Streams flow; C. Subterranean Reservoir filled with Water;—D. D. Borings in the Ground or Rock, called Artesian Wells.

25. ***But how Complete is the Design in*** this particular, also! The land is laid out by the hand of Providence, in channels and hollows, with streams, lakes, and reservoirs of water, on the ground, and under the ground, according to the plan which best contributes to the benefit of mankind.

26. ***By Boring or Drilling into the Earth,*** streams are met with at different depths, which are separated from each other by strata of rock; through the opening made, the water will rush upward as through a pipe, and rise like a fountain.

27. ***These Openings, or Borings, are called*** Artesian wells, from Artesium, now Artois, a province of France, where they have long been in use.

28. ***In many Places Water has been thus obtained*** in quantities sufficient for the working of heavy machinery.

29. ***In Dry and Desert Regions,*** even in Sahara, Artesian wells have been successfully sunk.

30. ***Some Artesian Wells have been sunk to Depths*** exceeding 2,000 feet, whence issues warm water; its temperature being derived from the internal heat of the earth.

31. ***In Wurtemburg,*** this water is introduced into pipes, for the heating of buildings, in winter; and by this means alone, the uniform temperature of 47° is maintained, while the temperature without is at zero.

32. ***At Paris,*** where the mean temperature at the surface, is 51°, the water of an Artesian well which is 1800 feet deep, has a constant temperature of 82°.

33. ***At St. Louis,*** the mean difference in temperature between the water obtained from an Artesian well, 1,500 feet deep, and that at the surface, is eighteen degrees.

34. ***At Charleston, S. C.,*** the temperature of the water at the surface averages 68°; at the depth of 500 feet it is 73°; at 1,000 feet, 84°; the average rate of increase of heat being about one degree for every 52 feet in depth.

35. ***Many such Wells,*** in New York, Pennsylvania, Virginia, and Ohio are famous for the quantities of salt and rock oil, or petroleum, obtained from them.

36. ***Petroleum has been collected,*** for centuries, in Birmah, Farther India, where it has been extensively used for producing artificial light; so, also, in northern Italy.

Section XIII.

RIVERS; THEIR SOURCES.

1. ***Rivers are Formed from Springs,*** or from rains that fail to penetrate the ground.

2. ***They commence*** as little streams, called ***Rills, or Rivulets,*** through which a child can wade, or over which he can step.

3. ***Always seeking the Lowlands,*** these rivulets meet other streams; and, enlarging as they go, soon become rivers.

4. ***Like a Dove set free,*** rivers seek their former home,—the ocean,—whether it be through extended plains, winding valleys, or mountain gaps. "Unto the place whence the rivers come, thither they return again." The dove seeks its home from a natural instinct; rivers seek the sea in obedience to the law of gravitation.

5. ***Some Rivers rise in Regions of great elevation,*** and at great distances from their mouths.

6. The ***Sources of the Amazon*** are far up the Andes; and, although they are within 100 miles of the Pacific, that river flows into the Atlantic, over a distance of about 4,000 miles.

7. ***Rivers are useful*** as great drains of the land; running off the surplus rain water into the ocean, and removing impurities from the surface of the ground. They also afford means of easy internal communication.

8. The ***Courses of Rivers*** are various, and are always governed by the slopes of the lowlands. Therefore, the general slopes of continents or countries can be determined from a common map, by the directions in which the rivers flow.

9. ***We Observe that nearly all the Rivers of South America*** flow in an easterly direction; hence, we know that the land east of the Andes, slopes towards the Atlantic.

Name the principal rivers of South America.

10. The ***Rivers of Northern Asia and Europe*** flow into the Arctic; hence, we know that from the Altai Mountains, the land presents a northern slope.

Mention the principal of these rivers. Mention the rivers of Eastern Asia, and the directions in which they flow. What is the slope of the land?

In what direction does the land of Southern Asia slope? Mention the largest rivers of the southern slope.

11. The land west of the Rocky Mountains slopes in what general direction? In what direction does the land of the United States, east of the Appalachian chain, slope? How do you ascertain this? The rivers of the United States, between the Rocky and Appalachian chains, flow into what river?

12. Mention the largest rivers on the western slope of the Mississippi basin; on the eastern slope.

What is the slope of the land of the Gulf States? Name the rivers of the southern slope.

How does the land slope in the region of Hudson Bay? Of the Baltic Sea? Of Western Africa?

13. The ***Sources of Rivers always occupy Higher Ground*** than do their mouths; many rivers, like the Ganges, have their sources several thousand feet above the level of their mouths, and owe their commencement to the melting snows of lofty mountains; consequently, their course to the sea is, at first, over very steep beds, or over a series of declivities, down which they plunge, producing rapids, cascades, and waterfalls. They approach their termination over beds less inclined, and comparatively level.

14. *Some Rivers*, like the Indus and Brahmaputra, flow for many miles on plateaus; others flow over beds of slight inclination from their sources to their mouths, and have no definite watershed. A boat may safely descend the Amazon River from the foot of the Andes Mountains to the Atlantic Ocean.

15. The *Waters of the Amazon are supplied* mainly by the excessive rains for which the Equatorial regions of South America are celebrated.

16. The *Upper Course of a River* commences at the watershed and continues over that part of its bed which is the most inclined: in this part, waterfalls and rapids are chiefly found.

17. The *Lower Course of a River* is toward its mouth; its bed is quite or almost level.

18. *By means of the Dissolving and Carrying Powers of Water*, the surface of the lowlands has received its comparative evenness.

19. The *Most Important River in North America* is the Mississippi. Its source is Itasca Lake, in the northern part of Minnesota, and is elevated nearly 1,700 feet above the level of the sea. Its general course is southward, and its total length about 3,000 miles.

20. The Mississippi is navigable by steamboats to the Falls of St. Anthony, 2,200 miles from the Gulf of Mexico; above the falls, it is also navigable for a considerable distance.

21. The *Mississippi and Ohio Rivers* constitute a line of communication between New Orleans and Pittsburg, of about 2,300 miles in length.

22. On the *Missouri River* steamboat navigation has reached to the foot of the Rocky Mountains, in the western part of Montana, a distance from the Gulf of Mexico of 4,000 miles.

23. The *Illinois River* is navigable for steamboats as far as La Salle, which is connected by a canal with Chicago, rendering navigation complete between Lake Michigan and the Mississippi, or between the Great Lakes and the Gulf of Mexico.

24. That the *Course of a River should not be in a Direct Line* to the sea, was wisely ordered by the Creator; for its various windings render the descent more gradual, and the current less rapid and destructive. Besides this, the winding course of a river increases the area of drainage, and the facilities for the progress of civilization and trade.

25. The *Distance from Cairo, Illinois, to New Orleans*, by the Mississippi River, is 1,178 miles. If there were no bends or windings in that river, the distance between these two places would be 700 miles less, but the force and destructiveness of the current would be greatly increased.

26. The *Niagara River*, on its way from Lake Erie to Lake Ontario, makes a total descent of 330 feet; about one-half of this descent is over a precipice, down which the waters rush with such tremendous force, that they are constantly wearing away the rocks beneath; therefore, Niagara Falls are gradually receding toward Lake Erie.

27. *Some Rivers of Mountainous Districts*, as in California, have worn long, narrow channels in the rocks, called *Cañons* (*kan'yuns*). A remarkable cañon of the Colorado River is in Utah and Arizona. Its length is 300 miles, and the rocks stand perpendicularly above the water in the gorge, to heights varying from 3,000 to 6,000 feet.

Passages worn through the Rocks on the Southern Coast of Norway.

28. *A River System* is composed of a river and its tributaries; thus resembling a great vine with its branches spread in all directions.

29. *A River Basin* comprises all the land that is drained by a river and its tributaries. In the lowest part of the valley flows the principal stream.

30. The *Basin of the Amazon* covers an area of more than 2,000,000 square miles; that of the Mississippi, about 1,000,000 square miles.

31. *A Watershed* is the ridge of land which surrounds a river basin and casts the water in different directions.

32. The *Watershed of Rivers flowing down opposite Sides of a Mountain Range*, is that part of the range which is elevated above the sources of those streams.

33. *A River Bed* is the ground over which the water flows. The channel is the deepest part of a river. The right bank is on your right hand as you sail down the stream; the left bank is on your left hand.

34. In many instances, *Springs but a few Rods distant from each other*, and fed from snows resting on the same peak, supply rivers which terminate at different sides of a continent.

35. The *Rivers of that Part of Europe* which is embraced by the Black, Mediterranean, and North Seas, have their sources among the mountains of Switzerland and very near together; yet they flow in different directions, and empty into seas lying on different sides of Europe.

Name the principal rivers which rise in or near Switzerland? Which flows east and into what does it empty? Which flows north? Into what does the Rhine flow? Which flows south? Into what does the Rhone flow? What river having some of its sources in the mountains of Switzerland flows into the Adriatic Sea.

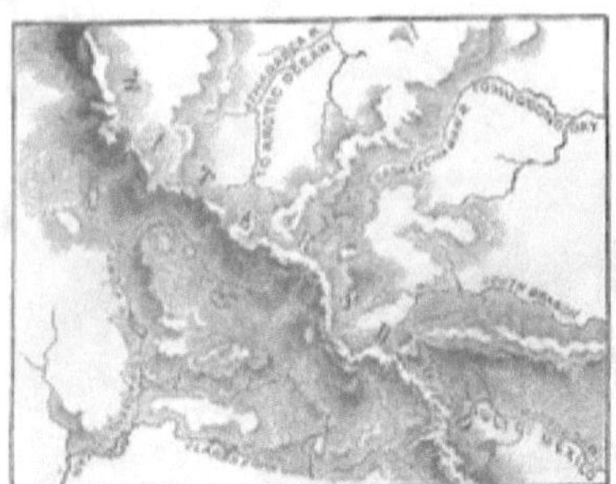

Watershed and Head-waters of Four Great River Basins in North America.

36. The *Head-waters of the Missouri and Clarke's Rivers,* in the Rocky Mountains, are almost together; yet the waters of one, by way of the Missouri and Mississippi Rivers, enter the Gulf of Mexico, and thence into the Atlantic; while the waters of the other, empty into the Columbia River and find their way into the Pacific.

37. *A Northern Tributary of the Columbia River* has its head-waters very near those of the Saskatch'ewan and Atha-bas'ca Rivers.

Where do these rivers rise? Into what does the Columbia River flow? The Missouri? The Mississippi? The Saskatchewan? The Athabasca? Which rise on the eastern slope of the Rocky Mountains? On the western slope?

38. *A House may be so located* upon the ridge which forms a watershed, that the rain falling upon one slope of its roof, may eventually find its way to one ocean, and that falling upon the opposite slope, to another ocean.

39. *Asia differs greatly from North America,* in this respect. The river basins of the Indian, Pacific, and Arctic slopes, are so disposed that the head-waters of their rivers are separated from each other by vast plateaus.

40. *Some Rivers do not empty into the Ocean,* but into an inland sea, or lake, as those of the Caspian Sea basin and the great basin of Utah. The river Jordan which flows into the Dead Sea belongs to this class.

41. *Some Rivers of Africa* disappear in the sands of the desert, and others are partly subterranean. These enter caverns, channels, or loose strata below the surface.

42. *Oceanic Rivers* are those whose waters reach the ocean, directly or indirectly; as the Amazon, Ohio, Danube, and Connecticut.

43. *Continental Rivers* are those of inland regions, whose waters do not reach the ocean; as the Volga and Ural.

44. *Many Rivers which have Rapid Currents* bear along with them alluvial washings from the land, and deposit them at their mouths, forming *deltas.*

45. The *Mississippi* and its tributaries are constantly transporting mud, logs, and stones, from the land of about twenty States and Territories, and depositing them in the valley of the Mississippi and at its delta.

46. *Borings have been made,* north of New Orleans, to the depth of 600 feet without reaching the bottom of the drifted mass; and, judging from the amount annually brought down by the Mississippi, it is estimated that the formation of land by its deposits, has already occupied more than 100,000 years. Hence, the land is constantly encroaching upon the Gulf of Mexico.

47. *This is also remarkable* in the Ganges, Nile, and Rhine.

48. Accordingly, the mountains and hills on the globe are being gradually diminished in height, and the land surface of the earth gradually extended.

Waterfall.—Saw Mill.

49. The *Streams rushing down the Mountain Sides,* are constantly carrying new soil to increase the fertility of the plains below. On their way down they turn the wheels of numerous mills and manufactories; and, by means of reservoirs and pipes, cities are abundantly supplied with fresh water.

50. The *Water of a River* is high or low, according as the season is rainy or dry.

51. *Many Rivers, like the Mississippi,* become full, sometimes to overflowing, by the melting of the snow at the approach of spring; but, during the summer months, the water is comparatively low.

52. *Other Rivers, like the Nile,* receive the tropical rains and rise periodically.

53. The *Sources of some Rivers,* like the Mackenzie and those of Siberia, are affected by the spring thaw, while their mouths, far northward, remain covered with ice; causing extensive overflows, by which stones, masses of earth, trees, and ice, are carried far across the land.

54. *Rivers opening into the Ocean* receive sea-water which is forced into them by the tides and winds; thus increasing their importance for purposes of navigation.

55. This is remarkable chiefly with rivers which open toward the east and south, owing to the westward movement of the tide.

56. The *United States and Europe* owe much of their greatness to their rivers, canals, and railroads, which intersect all their important parts.

57. *All that part of Europe lying West of the Black Sea,* is traversed by rivers which rise in the same region, and flow in all directions; while Asia and Africa contain immense tracts not crossed by a single river.

58. The *Importance of Rivers* to the development of mankind is manifested by the numerous villages and cities which line their banks; thus resembling the vine, whose value is indicated by the clusters of grapes hanging upon its branches.

59. Although South America is still in a backward state of development, its vast rivers and fertile plains promise it, in the future, a high rank among the divisions of the earth.

A Sectional View of the Great Lakes and the St. Lawrence River, looking North.

Section XIV.

LAKES; THEIR ELEVATIONS AND DEPTHS.

1. *Lakes* are collections of water in hollows of the land, of such a depth that their outlets cannot completely drain them.

2. *There are Four Classes of Lakes:*

3. The *First Class* has no streams which serve either as inlets or outlets.

4. The *Second Class* differs from the first in having an outlet; both classes are supplied by springs which burst forth from the bed of the lake.

5. The *Lakes of the Second Class* are generally situated on great elevations, and, in many instances, form the sources of rivers.

6. The *Third Class* both receives and discharges its waters by means of streams. Most lakes belong to this class.

7. The *Fourth Class* includes those lakes which receive streams of water, but have no visible outlet. They belong to continental or inland basins, and are numerous in Asia. These lakes are kept from overflowing their banks by means of evaporation.

8. *Many Depressions of the Land Surface* would contain lakes, but for the effect of evaporation.

9. *Nearly all Lakes* are supplied by streams which empty into them, and by springs rising from the bottom and sides.

10. *Some Lakes in Mountainous Regions* are supplied from the melting snow of the surrounding peaks.

11. *Lakes occur* in highlands and lowlands. Some are elevated several thousand feet above the sea level, while others are depressed below it.

12. The *Most Elevated Lake in the World*, is Lake Sir-i-kol, which is situated on the mountains in the western part of the Chinese Empire. It is about 15,000 feet above the level of the sea. (*See Illustration* on page 17.)

13. *Lake Titicaca*, between Peru and Bolivia, is over 12,000 feet above the level of the ocean. Its area is more than 2,000 square miles, and its depth is equal to that of Lake Ontario.

14. The *Dead Sea*, properly a lake, is more than 1,300 feet below the sea level. It is the greatest depression of the kind on the globe. This famous lake, whose formation resulted from the catastrophe which destroyed the cities of Sodom and Gomorrah, about 1,900 B. C., contains a for greater portion of salt than do other salt lakes; the water being so impregnated with it, that even heavy bodies float buoyantly. Asphaltum, in large quantities, and sulphur, are found on its banks.

15. The *Waters of most Lakes are Fresh;* but those having no outlet are usually salt. This is because all streams receive from the land through which they flow, small quantities of salt, which the waters hold in solution until it reaches the ocean, or another body of water having no outlet; here the salt is deposited.

16. The *Most Celebrated Salt Lakes* are the Caspian Sea, Aral Sea, and Dead Sea, and the great Salt Lake of Utah.

17. The *Basin of a Lake* comprises all the land drained by the streams which flow into the lake. It may be seen on a map, by passing a line around the sources of all its tributaries.

18. *Subterranean Lakes* are numerous. They are collections of rain water in caverns which are below the surface of the ground.

19. *Subterranean Lakes and Streams* frequently cause destructive inundations. The water and steam thrown up by volcanoes proceed from these lakes.

20. The *Island of Trinidad*, situated near the mouths of the Orinoco River, contains a lake three miles in circumference, that is famous for the quantities of pitch contained in its waters. This substance, like petroleum, is raised by the agency of subterranean fire.

21. *Lake Superior* is the largest body of fresh water on the globe. Its area is 32,000 square miles, and is equal to about three-fifths that of England.

22. Are the waters of the Great Lakes salt, or fresh? What river forms their outlet? In what direction does the St. Lawrence River flow? Mention the depth of each lake. Which is the deepest,—the shallowest? Which has the most elevated surface? Between what two lakes are the Falls of Niagara situated? From which does the water of the falls proceed? In what part of the St. Lawrence are the Thousand Islands? The Rapids?

Section XV.

The Atmosphere;—The Winds.

1. The *Atmosphere* is a gaseous fluid which surrounds and rests upon the earth.

2. *It is as necessary to Life*, as are water and food; neither plants nor animals could exist without it.

3. *Air consists of Two Gases*, oxygen and nitrogen, mixed together.

4. The *Ingredient of the Air which sustains Animal Life*, is oxygen, but, should these two gases be separated, the result would be instant death.

5. The *Air, like Wholesome Food*, is necessarily composed of both nutritious and innutritious substances.

6. *Oxygen forms* about one-fourth of the air; nitrogen, three-fourths.

7. The *Weight of the Atmosphere* is about $\frac{1}{833}$ that of water.

8. *It is Heaviest* at the surface of the earth, and diminishes in density, according to the distance above the surface.

9. *On the Tops of the Highest Mountains*, the air is so thin that man cannot breathe there.

10. The *Atmosphere extends upward*, to a distance, it is supposed, of about fifty miles.

11. *Winds* are currents or movements of the air, caused by the different degrees of temperature to which the air is subjected, and by the revolution of the earth upon its axis.

12. The *Air is Warmed*, partly by the passage through it of the sun's rays, but mostly by the radiation of the sun's heat from the earth's surface; consequently, the warmest part of the atmosphere is that which is in contact with the surface of the earth.

13. The *Heat and Density of the Atmosphere* diminish according as the elevation is increased; this has been observed by travelers who have ascended high mountains, and by æronauts in their balloon ascensions.

14. *As that part of the Atmosphere is Warmest which is nearest the Surface*, the upper and surrounding cold air presses down and replaces the warm and light air, which rises to more elevated regions.

15. *A Balloon ascends because* it is filled with a gas that is lighter than common air. When the gas is allowed to escape, the surrounding air rushes in and causes the balloon to descend.

16. The *Two General Movements of the Air* are from the Equator to the Poles, and from the Poles to the Equator.

17. *As the Cool and Heavy Winds press toward the Equator*, they are unable to keep up with the *eastward* motion of the Equatorial regions of the earth; and, by falling behind, they appear as a current of air moving *westward*.

(*For further explanation, see page 23, paragraphs 6 and 7.*)

18. *A Current of Water* receives the name of the direction *toward* which it flows; but a current of air, that *from* which it moves. Therefore, a *westerly* current of water and an *east* wind move in the same direction.

19. *Changes in the Courses of the Winds* are caused by various bodies of land, and by high mountain ranges.

20. *Where Two Winds from Different Directions meet*, they counteract each other's force, and cause calms; hence, there are Equatorial Calms, Calms of Cancer, Calms of Capricorn, and Polar Calms.

21. The *Trade Winds* of the Northern Hemisphere blowing from the north-east, and those of the Southern Hemisphere blowing from the south-east, meet near the Equator, and neutralize each other; thus causing calms in that region around the earth.

22. The *Winds* then rise to a greater elevation and tend toward the North and South Poles, moving over the tropical regions as *upper* currents.

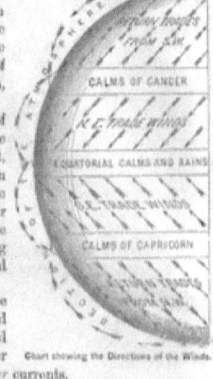

Chart showing the Directions of the Winds.

23. When they reach the temperate latitudes, they have become so cool and heavy on account of their elevation, that they descend to the surface, and blow from the south-west in the Northern Hemisphere, and from the north-west in the Southern Hemisphere. These are called *the Return Trades or Passage winds.* (*See page 24, paragraphs 10 to 14 inclusive.*)

24. The *General Direction of the Winds in the Tropical Regions* is toward the west. These winds contribute to the westward flow of the Equatorial Current.

25. *In the Temperate Regions* there is a like correspondence between the Return Currents of the ocean and the Return Trade Winds; their motion being toward the east.

26. *A Voyage from the United States to England*, in a sailing vessel, is made several days shorter by the aid of these winds and the Gulf Stream, than that from England to the United States. The time made by steamers from New York to Liverpool, is between nine and twelve days; but, returning, they require from two to four days longer.

27. The *Prevailing South-west Winds of the North Temperate Zone*, passing over the warm waters of the Gulf Stream, contribute largely to the advantages of Western and Southern Europe in climate, productions, and general development.

28. *If the Earth revolved on its Axis in the Opposite Direction*—from east to west—in what direction would the Trade Winds and the Equatorial Current move? If the earth did not revolve on its axis, what would become of the ocean currents and the winds?

29. The *Plan of the Winds*, like that of the ocean currents, is such that a constant circulation of air is maintained between the Eastern and Western Hemispheres, round and round the globe, and between the Northern and Southern Hemispheres, from the burning zone of the Tropics to the frozen regions of the Poles. "The wind goeth toward the south, and turneth about unto the north; it whirleth about continually."

Sea Breeze. A View on the Sea Coast. From Morning until Evening the Air which is over the Sea is Heavier than that over the Land; consequently, the Wind blows all Day from the Sea.

30. The ***Plan of Differences in Nature***, producing contact, opposition, and variety, is beneficial to mankind.

31. ***It is recognized*** in the light of day and the darkness of night, in land and water, sunshine and rain, in the variety of productions, and in the diverse pursuits of people.

32. The ***Wisdom of this Plan*** appears not only in the existence of such differences, but in their *coöperation* and *unity*.

33. ***Sea Coasts and Islands*** enjoy a more even temperature throughout the year than inland districts, because the ocean does not change its temperature, either in summer or winter, so readily as the land.

34. Winds which blow over the sea are generally not so cold in winter, nor so warm in summer, as those blowing over the land.

35. ***Land near the Sea is Warmer during the Day*** than the neighboring water. Sea air is then cooler and heavier than the air of the land; hence, the wind blows all day from the sea, and is called a *sea breeze*.

36. ***After Sundown***, as the land becomes cooler than the water, the air rushes back from the land, and is called a *land breeze*.

Land Breeze. At Night, the Air which is over the Land becomes Heavier than that over the Water, causing the Wind to blow all Night from the Land.

37. ***Land and Sea Breezes*** **are winds which** blow alternately from the land and sea.

38. They occur **on coasts and in** islands, especially in the tropical regions; **also on the** shores of large lakes.

39. ***In the Northern Hemisphere***, a north wind is cold, and a **south** wind, **warm; in the** Southern Hemisphere, the north wind is **warm, and the south,** cold.

Section XVI.

Moisture in the Atmosphere.

1. By heat, ***Water is Expanded*** and made lighter than the **air.**

2. The ***Water then rises*** in the form of vapor, and is carried away by the winds.

3. ***Vapor when Influenced by a Cool Temperature*** becomes condensed, and returns to the surface of the earth in the form of rain, snow, and dew.

4. The ***Motive Power of the Steam Engine*** is due to the property which water possesses of being easily expanded by heat and condensed by cold, thus forming a vacuum.

5. ***As the Air becomes Warm***, its capacity of holding moisture increases, and as the temperature falls that capacity diminishes. This difference between the temperature of the day and that of the night, causes dew to appear upon the grass and flowers, that they may be refreshed in the absence of rain.

6. ***Trees and Plants*** obtain much of their nourishment from the moisture in the air which is condensed by means of their leaves.

7. ***Vapor is not always Visible***, because it is spread out in the atmosphere, like the moisture that is exhaled in breathing. A pitcher of cold water placed in a warm room condenses vapor, which appears on the surface in the form of drops.

8. ***Evaporation increases*** with the warmth and dryness of the atmosphere; hence, the amount of rain is greatest in the tropical regions, and diminishes toward the poles.

9. ***Evaporation modifies Temperature.*** Without evaporation, the surface of the ocean would become hotter and hotter by the influence of the sun, and would therefore greatly intensify the heat of the atmosphere in contact with it. But not thus defective are nature's laws.

10. ***As Water becomes Heated at the Surface***, it gives place to cooler portions beneath, by rising, in the form of vapor, into upper and cooler regions of the atmosphere.

11. ***By the Action of the Waves***, lower and cooler portions of the water are brought up to the surface to reduce its temperature.

12. ***By these Movements of the Water***, the surface of the ocean is prevented from attaining a degree of heat so great as to prove detrimental to the comfort and interests of mankind.

13. ***On the Land***, these two movements do not occur. Its heated surface cannot rise in the air as water does by the process of evaporation; neither are cool portions of the ground brought constantly up to reduce the temperature of the surface; hence, the land becomes more heated by the sun's rays than the water does.

14. ***In Summer***, the land freely imparts its heat to the atmosphere near it and makes that season hot, perhaps oppressive; but when winter comes, the land has not saved enough heat to keep off the severity of the cold. It is owing to this process of radiation that in some inland places, hot and sultry **days are followed** by chilly and disagreeable nights, and that **the deposition of** dew is greater **on** land **than on** water.

View from the Catskill Mountain House, New York, looking East. The Hudson River appears in the Distance.
Names of the Classes of Clouds:—1, Cirrus; 2, Stratus; 3, Cumulus; 4, Nimbus.

15. *In Winter,* the continental climate is colder than the oceanic, because the land parts with its heat by radiation more readily than does the water.

16. *St. Petersburg and the Faroe Islands* are nearly in the same latitude: the climate of the former is *continental*; of the latter, *oceanic.* Which is the warmer in summer? Which is the colder in winter?

17. The *Summer of St. Petersburg* averages seven degrees warmer than that of the Faroe Islands, north-west of Scotland; while the winter of the former is twenty-two degrees colder than that of the latter.

18. *Clouds* are collections of visible vapor suspended in the atmosphere, at altitudes ranging from one to five miles.

19. *Fog* is a like collection nearer the earth's surface.

20. *Vapor* consists of particles of water so fine and light that they float in the air like dust.

21. *There are Four Classes of Clouds:*

22. The *Cirrus,* which is the highest cloud we see, is of a light feathery form; and, on account of its elevation, its vapors probably exist in light particles of snow.

23. The *Stra'tus* exists generally in the night and in winter; it is formed by the cooling and consequent settling down of the higher clouds, which appear in horizontal bands.

24. The *Cu'mulus* is the summer-day cloud which forms at sunrise by the gathering together of the night mists.

25. The *Nimbus* is the heavy, dark cloud from which rain falls.

26. *When Clouds pass into the Atmosphere* [illegible] rounds the cold summits of the mountains, their v[illegible] come condensed, and fall in the form of rain and sno[illegible] which supply springs, streams, and lakes of elevated regions.

27. *Rain falls from Clouds* at different elevations; in mountainous districts heavy showers sometimes fall upon the low ground, while persons on a mountain behold a clear sky above them and black clouds below them.

28. *If there were no Mountains* on the globe, the clouds would pass over the land without depositing an amount of rain sufficient for the preservation of vegetable and animal life.

29. The *Harmony which exists* between the influence of the mountains and the movements of the clouds, produces results necessary to the development of the earth and to the well-being of man. Is this harmony the result of accident, or is it in accordance with the wise design of the Creator?

30. *Rain* is caused by vapor entering a cool atmosphere and becoming condensed; it then falls to the earth in drops.

31. *If Rain, in its Descent, passes through a Current of Air* sufficiently cold to freeze the drops, hail is produced.

32. *If Vapor becomes Frozen* while its particles are light, it falls to the earth in the form of snow.

33. *In North America,* snow is seldom seen to fall south of the parallel of latitude 30°—that which passes over New Orleans.

34. *In the Hot Zone of South America,* however, it remains throughout the year on all mountain peaks above the elevation of 16,000 feet.

35. *Snow is a Non-conductor of Heat;* it consequently prevents radiation of heat from the ground covered by it, and protects roots, vegetables, and seeds from the intense cold of winter.

36. *Rain is distributed* over the land by the agency of winds.

37. The *Chief Source of Supply* is the ocean; although from every lake, pond, and stream, there arises moisture which returns to refresh vegetation.

38. The *Greatest Amount of Rain* falls within a belt around the earth, near the Equator. This is because the Trade Winds here come in contact with each other and carry the vapors with which they are heavily charged, up into a cool atmosphere which condenses them.

39. *On the Continents,* the greatest amount of rain falls near the sea coasts and upon the mountainous regions in the interior.

40. *On the Western Continent,* the greatest amount of rain falls in South America—on its eastern coast, and the eastern slope of the Andes Mountains.

41. *Ranges of Mountains,* like the Andes, whose tops are perpetually covered with snow, cause vast quantities of rain to fall on the windward side or slope, while in some places on the opposite slope, rain is almost or wholly unknown.

42. The *Desert of Atacama* (*ah-tah-kah'mah*) is situated west of the Andes, and lies partly in Peru and partly in Bolivia. The east winds are deprived of their moisture before passing the mountains, and continuing westward, prevent the vapors of the Pacific from reaching that arid region. (See Sec. IX., par. 14.

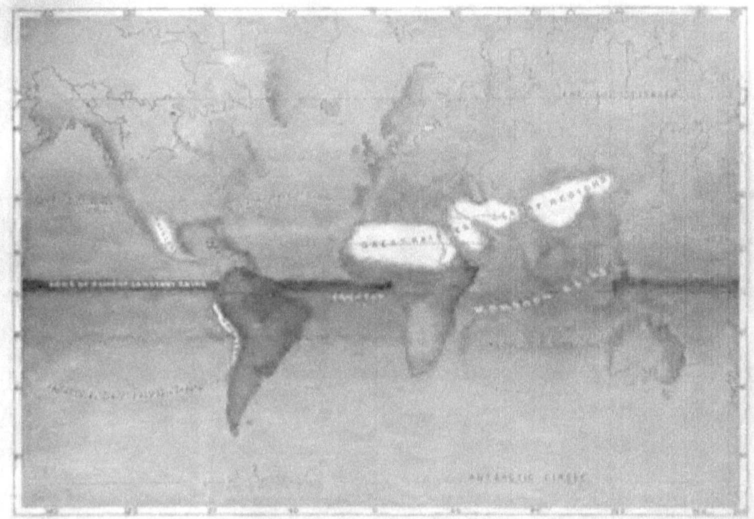

Rain Chart.—The Quantity of Rain which falls at any Place is indicated on this Chart by the Depth of the Shading; the Darker the Shading, so much Greater is the Amount of Rain.

43. The ***Rainless Region of South America*** lies west of the Andes, and in the track of the *South-east Trade Winds.*

44. ***South of the Desert of Atacama*** is Chili, which, although lying on the west of the Andes, receives copious rains; these are brought by the *Return Trade Winds*, which blow over the Pacific from the north-west.

45. The ***North-western Coast of South America*** is within the zone of almost constant rains.

46. The ***Trade Winds*** blow from the east and deposit their rain mostly on the eastern coasts of continents and islands, and on the eastern slopes of high mountain ranges.

47. The ***Return Trades*** deposit their rain chiefly on western coasts and slopes.

48. ***In North America***, rain is most abundant on its western side, and around the Gulf of Mexico.

49. The ***West Indies are Noted*** for the heavy rains which fall there; these rains proceed from the vapors supplied by the warm waters of the Gulf Stream.

50. The ***British Islands***, together with the western and southern coasts of Europe, are supplied with rain from the vapors of the Atlantic Ocean, which are carried there by the prevailing west winds—the Return Trades; while on the plains of Russia and Siberia, the amount of rain is comparatively slight.

51. The ***Rains of Africa***, like those of South America, are supplied by means of the Trade Winds; but while South America has its high mountain range on its western side, causing copious rains to fall upon vast plains eastward, the high mountains on the eastern side of Africa return much of the rain immediately into the Indian Ocean.

52. *Do the Trade Winds deposit more Rain* on the eastern, or the western sides of islands and mountains? On which coast of South America is rain most abundant? On which coast of Africa? On which side of the Andes Mountains? On which side of continents, islands, and mountain ranges do the *Return Trades* deposit most rain?

53. *Do Vapors rise mostly from Cold, or Warm Currents?* From what current do vapors come which supply the rivers of Western and Southern Europe? From what current are the rivers of the Pacific coast of North America supplied? (*See Chart on page 23.*)

54. *On which Coast of Greenland is Rain most Abundant?* On which side of Norway? France? Spain? Arabia? Australia? Hudson Bay?

55. *What Great River in Africa* flows through the rainless district? Whence does the Nile receive its waters?

56. ***If no Ocean intervened between America and Europe***, the absence of rain alone would make Europe desolate.

57. The ***Great Rainless Region of the Old World*** includes the Great Desert of Africa and the deserts of Arabia, Persia, and Cobi.

58. ***Their Condition is caused***, mainly, by their interior position, the comparative dryness of the winds, and the absence of lofty peaks that would act as condensers of the thin and scattered vapor which floats over them.

59. ***In the New World***, the principal rainless districts are in Mexico and Central America, and in South America, on the western side of the Andes.

60. ***In some Places where Rain seldom or never falls***, vegetation is sustained by frequent and heavy dews.

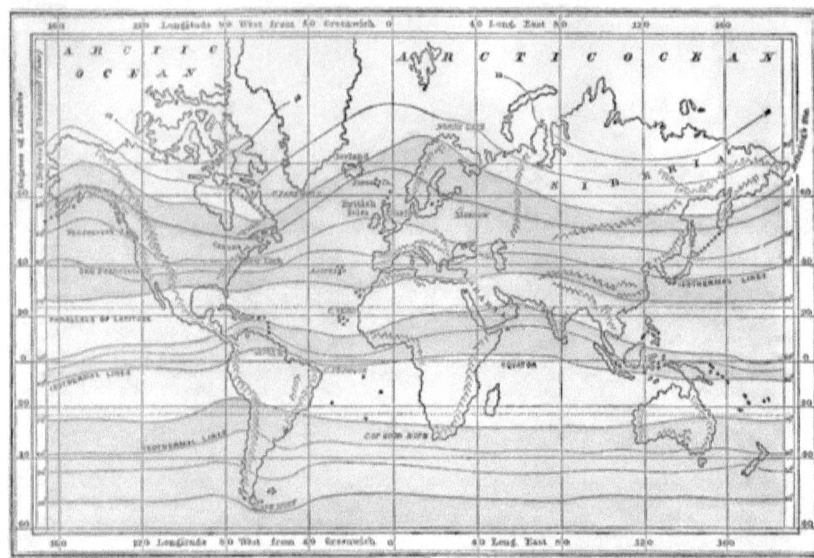

Chart showing Isothermal Zones and the Mean Annual Temperature of the Different Parts of the Earth's Surface.

Section XVII.

Climate,—Isothermal Lines.

1. ***Climate*** is the condition of a place in relation, chiefly, to the temperature and moisture of the atmosphere.

2. ***Isotherms, or Isothermal Lines,*** are lines drawn on a a chart through places of equal mean temperature.

3. ***Mean Annual Temperature*** is midway between the heat of summer and the cold of winter. In Cincinnati the mean temperature of summer is 73°, and of winter, 33°; the mean annual temperature is 53°, which is obtained thus:

$$\frac{73° + 33°}{2} = 53°.$$

4. ***If the Temperature diminished uniformly*** from the Equator to the poles, isothermal lines would correspond with parallels of latitude.

5. ***Their Directions*** are various, and indicate the influence upon climate, of ocean currents, winds, high mountains, frozen plains, and burning deserts.

6. Therefore, ***the Hot, Cold, and Temperate Zones*** of the earth are situated between isothermal lines, and not between parallels of latitude. These zones are called *Isothermal Zones.*

7. ***Isothermal Lines have their Greatest Inclination*** in the North Atlantic Ocean, and show that the north-west coasts of the Old World possess warmer climates throughout the year than other parts of the land, at the same latitude.

8. ***This is chiefly owing*** to the influence of the Gulf Stream, which warms the prevailing south-west winds passing over it on their way toward the west coasts of Europe.

9. ***Eastward from these Coasts,*** the temperature gradually falls, as shown by the isotherms, on account of the cooling influence of the high mountains of Europe and Asia, and the frozen plains of Siberia.

10. ***If the Waters of the Atlantic imparted no Warmth*** to the atmosphere, Newfoundland and Northern France, being between the same parallels of latitude, would have the same climate.

11. ***Without the Influence of the Gulf Stream,*** the now genial and productive climate of the British Isles would be similar to that of the cold and desolate regions of Labrador.

12. In reality, however, the ***Center of Great Britain,*** at the latitude of 55°, has the same mean temperature as the eastern side of the United States, at the latitude of 40°.

13. The ***Isotherm which passes through Newfoundland*** extends north-eastward to the coast of Iceland, 15° nearer the North Pole.

14. The ***Temperature*** of the coast of Norway is the same as that of Central Labrador, although 20° of latitude lie between them. The influence of the Gulf Stream is felt upon the coasts of Spitzbergen and also upon the north coast of Nova Zembla.

15. The *Land of the Northern Hemisphere* may be divided into six climatic zones: The Torrid or Hottest, the Hot, Warm, Temperate, Cold, and Frigid or Coldest.

16. THE MEAN ANNUAL TEMPERATURE OF THE ZONES.

The Frigid Zone, below	32°	Fahr.
The Cold Zone, between	32° and 40°	"
The Temperate Zone, between	40° and 60°	"
The Warm Zone, between	60° and 70°	"
The Hot Zone, between	70° and 80°	"
The Torrid Zone is over	80°	"

17. The *Isotherm of 32° Fahr.* is the line of constantly frozen ground.

18. *Through what Parts of North America does the Isotherm of 32° pass?* Through what parts of Europe? Of Asia? What large bay in British America receives cold water from the Arctic Ocean? What effect has the temperature of the water of Hudson Bay upon the climate of the surrounding regions? What is the direction of the isotherms which pass over those regions?

19. *What Places are under the same Isotherm as New York?* What is their mean temperature?

What places are under the isotherm which passes over Panama? What is their mean temperature?

What places are under the isotherm which passes over Newfoundland?

20. *What Parts of the Northern Hemisphere are in the Hottest Zone?* The Hot Zone? The Warm Zone? The Temperate Zone? The Cold Zone? The Frigid Zone?

What lands of the Southern Hemisphere are in the Hottest Zone? The Hot Zone? The Warm Zone? The Temperate Zone?

Does any part of the two continents extend south of the line of constantly frozen ground? What part extends furthest south?

What is the mean annual temperature of Cape Horn?

21. The *Prevailing Winds of the United States and Europe* blow from the south-west; consequently, they are *Land Winds*, to the eastern parts of the United States and Europe, and cause *Excessive Climates* (see page 33, paragraph 34); while to the western coasts, they are *Sea Winds*, and produce that evenness of climate for which Western Europe and the Pacific coast of the United States are remarkable.

22. *If we Compare the Climate of New York* with that of San Francisco, the difference between oceanic and land climates will be obvious.

THE MEAN TEMPERATURE OF THE HOTTEST AND COLDEST MONTHS DURING THE YEAR, IN NEW YORK AND SAN FRANCISCO.

Hottest month in New York,	88° Fahr.;—San Francisco,	59°.
Coldest " " " "	33° " " "	50°.
Mean difference between summer and winter	55° " " "	9°.

23. *While Snow usually lies in New York a great part of the Winter*, it rarely falls in San Francisco. The winter of San Francisco consists of a *Rainy Season*, which is caused by the cooling influence of the mountains upon the moisture of the sea winds. Its summer is known as the *Dry Season*.

24. The *Temperature of the East Coast of the United States* is farther depressed by cold waters from the Arctic Currents, which here flow in a south-westerly direction between the Gulf Stream and the coast. It is therefore a counter current.

25. The *Valleys near the Coast of California* possess a more even and delightful climate than any other part of the world.

26. *In some parts of the Faroe Islands*, water never fr[eezes]; while in Yakoutsk, a city of Siberia, which lies under [the] same parallel, the summers average 9° warmer, and the winters, 76° colder. The mean difference in temperature between summer and winter at the former place, is only 15°; at the latter, it is 100°.

27. *In which of the two Places* just mentioned is the climate excessive? Even? Continental? Oceanic?

28. *In the Azores and Madeiras*—islands north-west of Africa,—the climate is that of eternal spring; flowers bloom there throughout the year in the open air, although those islands are between the same parallels as Philadelphia, Cincinnati, and St. Louis.

29. *Forests, Fertile Plains, and Parched Deserts* owe their respective conditions not only to their position on the globe, but also to the influence of ocean currents, the agency of winds, and the presence or absence of rain.

30. The *Isotherms of North America, Europe, and Asia* extend in the same general direction—south-eastward from their western sides; showing the mean temperature of their western coasts to be warmer than that of their eastern.

31. The *Climate of the Atlantic Coasts of Europe* corresponds with that of the Pacific Coast of North America.

32. Isothermal lines correspond more nearly with parallels of latitude in the Water Hemisphere than in the Land Hemisphere, showing the evenness of an oceanic climate.

33. *Compare the Climate of Vancouver's Island with that of Maine.* In the former, the summers are mild, and the frosts of short duration; while in the latter, the summers are hot, and the winters very severe, the snow lying on the ground from three to five months in the year.

34. *Traveling Eastwardly from the Pacific Coast* of North America on any parallel north of San Francisco, what change of temperature is observed? (*See Isothermal Lines.*)

35. *Sailing Due East from the Atlantic Coast*, what change?

What part of the Pacific coast of North America has the same temperature as Newfoundland? Give the latitude of each of these two places. What is the average temperature?

What island on the Pacific coast of North America has the same temperature as New York? What is the latitude of each? Their mean temperature?

36. *What European Country* has a climate similar to that of California?

Although North Cape is 11° further north than Cape Farewell, its climate is no colder. Why?

What city in Russia has the same latitude as Glasgow? At which place is the winter more severe? Why?

37. *Why does the Climate of the West Indies* differ from that of Newfoundland?

Which is farther north—Canada, or Ireland? In which are the winters more severe? Why?

Which coast of the United States possesses the more even climate—the Atlantic, or Pacific? Why?

38. The *Climate of the Western Side of North America* and of Western Europe is more conducive to health than that of their eastern parts, on account of its greater evenness.

39. *If the Bed of the Atlantic should be elevated* and become dry land, what climates would be affected, and how?

If a range of high mountains extended along the west coast of Europe, what would be the effect upon the climate and rains of that division?

40. *Why is the Climate of the Atlantic Coast* of North America warmer in summer, and colder in winter, than that of the Pacific coast?

What effect have the Rocky Mountains upon the temperature of the westerly winds of the United States?

A Mountain Stream.

41. *Activity, Use, and Influence* are everywhere, from the mighty ocean and lofty mountains to the little stream that turns the miller's wheel and furnishes drink to cattle.

42. The *Common Garden Worm* opens channels in the ground through which the moisture enters to nourish the roots of plants, and otherwise assists man in preparing the soil.

43. The *Ocean,* although covering the greater part of the earth's surface, is not a vast waste, for it supplies the land with vegetation and an abundance of fresh water for the support of all life; and, as the modifier of climate, it exerts its essential influence upon the physical, intellectual, and moral conditions of mankind, and contributes largely to the prosperity of the nations of the earth.

The Earth in the form of a Globe.

The Earth in the form of a Cube.

The Earth in the form of a Cylinder.

44. *None can fail to recognize the Systems of Winds* and ocean currents as necessary to the life and well-being of the earth's inhabitants; and, herein, the wisdom of the plan by which the world was made in the form of a *globe.*

45. *If the World had been made in the Form of a Cube,* or of a cylinder, there would not be that harmony of action between diverse conditions of the earth's surface which now exists.

If the Earth were a Great Cube, would there be zones of different degrees of temperature as there are now? The same winds and ocean currents?

46. *Diversity in Climate and Productions* of the earth, and in the pursuits of individuals and nations, constitutes a wise provision of the Creator.

47. *All the Great Agents* by which the various conditions of the earth are so wonderfully sustained, are so adapted to each other, and act together so harmoniously, that if but one should neglect to act its part, mankind would suffer—perhaps perish.

48. *If the Process of Evaporation* should be discontinued, what would be the effect upon vegetation, animals, and man? Or, if all winds should cease, where would all the rain fall?

49. The *Southern Part of the United States* is admirably adapted to agriculture. Its peculiarities of soil and climate so harmonize with each other that the amount of cotton alone which is here produced, and upon which millions of the earth's inhabitants—on both continents—depend for clothing, comprises nearly seven-eighths of the entire yield of the world.

50. The *Rugged North-eastern Part of this Country* is provided with coal, iron, and mountain streams, which make it the great manufacturing region of the Union.

51. *If the Gulf and Atlantic States of the South* were mountainous, and the north-eastern States level, the cotton plant, sugar-cane, and rice would not grow either upon mountains of the south or cool plains of the north-east.

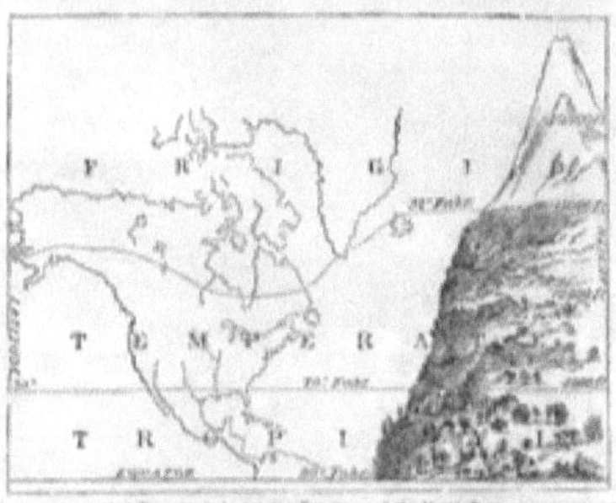

Chart, showing that Climates between the Equator and the North Pole correspond with those on the Sides of High Mountains at the Equator.

52. *Temperature so diminishes with Increase of Elevation* that various climates, with their characteristic productions, are found not only upon the earth's surface between the Equator and the Poles, but likewise upon the sides of high mountains between their base and summit.

53. *If we consider the Northern Hemisphere* and the side of a mountain which is situated under the Equator, to be divided each into three climatic zones, the Torrid Zone on the former would extend *northward* to about the parallel of 30°, and on the latter, *upward* to the elevation of about 5,000 feet; the Temperate Zone of the former would extend to about the Isotherm averaging 60° latitude, and on the latter, to the height of about 15,000 feet.

What part of the earth's surface and what part of a tropical mountain have a mean temperature of 80° Fahr.? Of 70°? Of 34°?

54. *From the Equator toward the North Pole,* the temperature diminishes about 1° for every 100 miles.

55. *From the Level of the Ocean* to the summit of a mountain, the temperature diminishes about 1° for every 350 feet.

FROZEN REGIONS.

56. The *Upper Part of this Picture* represents the regions of perpetual snow among the tropical Andes, which correspond, in temperature, to the Frigid Zones.

These High Snow-clad Peaks are the great condensers which bring down moisture from the atmosphere, and supply the rains which fill the lakes and rivers of South America.

TEMPERATE REGIONS.

57. The *Middle Portion* of the picture represents a region whose climate corresponds to that of the Temperate Zones.

This Region contains plateaus and elevated cities, whose inhabitants enjoy a cool and salubrious climate.

Depressions on the surface of the plateaus form the beds of elevated lakes and streams, which receive their waters from the melting snows above them.

Here are Fertile Fields of grain and grass; here flourish trees, fruits, and plants peculiar to the Temperate Zone.

TROPICAL REGIONS.

58. *Below the Line* which marks an elevation of 5,000 feet above the level of the sea, is the climate which corresponds to that of the hot zone of the earth, not only in temperature, but also in its vegetable productions and species of animals.

At various Heights, are deep ravines and fearful precipices, down which rush streams and waterfalls.

View among the Andes Mountains, showing that different Zones of Temperature pertain to different Elevations.

FROZEN REGIONS.

59. The *Highest Peaks* of the Tropical Andes are elevated above the level of the sea about 20,000 feet.

The *Most Noted* are Chimborazo, Sorata, Illimani, Antisana, Cotopaxi, and Arequipa.

An immense bird, called the condor, builds its nest far up these heights, and has been known to fly above the summit of Chimborazo.

TEMPERATE REGIONS.

60. The *City of Potosi* is represented on the right of the illustration. It is built on a plateau, at an elevation of more than 12,000 feet above the level of the sea, and contains about 30,000 inhabitants.

Quito is represented on the left, at an elevation of about 10,000 feet, and, although almost immediately under the Equator, its temperature is that of continual spring.

Surrounded by plains and fertile valleys which are enclosed by lofty mountains, Quito is celebrated for the grandeur of its scenery.

TROPICAL REGIONS.

61. *At the Foot* of these mountains the heat is oppressive throughout the year.

The *Trees* of the lower or hot section comprise the palm, tree fern, banana, and pine-apple.

The Animals comprise the tapir, jaguar, cougar, and several tribes of monkeys; besides, parrots, macaws, and other birds which are noted for the brilliant colors of their plumage.

62. *A Traveler ascending a High Mountain* of the tropical Andes, passes through climates similar to those of the different zones, from the heat of the Equatorial, to the continual frost of the Arctic regions.

63. *At the Base of the Mountain*, or at the ocean level, he endures the oppressive heat of the tropical sun, and observes the luxuriant vegetation, lofty trees, and luscious fruits of the *hot zone*.

64. *Half-way up the Mountain*, he enjoys the delightful air of the *Temperate Zone*, with its characteristic varieties of trees, plants, and grains.

65. *Continuing to ascend*, he observes that the mercury in the thermometer is gradually falling, and passes through regions whose temperature admits only of the growth of low evergreens, stunted shrubs, and mosses.

66. *As the Traveler approaches the Top*, he enters the region of perpetual snow, and experiences a climate similar to to that of the Esquimau or the Laplander.

Section XVIII.

VEGETATION; ITS GROWTH AND USES.

1. *From Vegetation*, all animal life derives its food, either directly or indirectly. Some animals subsist on flesh, which, however, is the flesh of animals that have fed on vegetation.

2. *For this Reason*, the Creator has covered the greater part of the land with vegetation; for this reason, He made the grass, herbs, and trees, before living creatures were brought into existence.

"He causeth the grass to grow for the cattle, and herb for the service of man."

3. The *Inhabitants of One Climate* require food different from that required by the inhabitants of another climate.

4. *Differences in Temperature*, soil, and degree of moisture on the earth's surface, produce differences in the kinds of plants, and furnish to the various races of mankind and species of animals, the food which is best suited to their wants.

5. The *Inhabitants of the Hot Zone* require food of a light or watery nature; therefore, that region is provided with abundant and luscious fruits, besides rice, millet, and sago.

6. *When you leave the Tropical Regions* and enter a cooler climate, food of a more substantial nature is required.

7. *In the Temperate Zones*, food is obtained mainly from the heavier grains and the flesh of animals.

8. *In the Frigid Zones*, the inhabitants subsist almost entirely on animal food.

9. It is therefore, *according to a Wise Design* that the tropical regions yield the most abundant vegetation.

10. The *Conditions which are most favorable* to the growth of plants, are heat and moisture.

11. *Trees supply Man with Ripe Fruits* and afford shelter during the hot season; some are cut down and sawed into lumber for building purposes and for fuel.

12. *From Plants*, man obtains food for himself and for the animals which are useful to him.

13. The *Most Important Food Plants* are wheat, corn, rice, oats, rye, and potatoes.

14. *Plants derive their Nourishment* from the water which they receive from the soil through their roots, and from the atmosphere through their leaves.

15. *Plants are provided* with cells or tubes through which the water circulates. Those plants which have the largest cells, roots, and leaves, require most water.

16. *Water holds in Solution* various substances that are contained in the soil and are required for the growth of plants; these are, chiefly, carbonic acid, with animal, vegetable, and earthy substances.

17. *Carbonic Acid Gas* is exhaled from the lungs of animals; and, although poisonous to all living creatures, it furnishes the material which enters largely into the formation of trees, vegetables, and flowers.

18. *Herein is the Economy of Nature* plainly manifested; vegetation sustains animal life; animal life and animal substances sustain vegetation. They depend upon each other.

19. *Vegetation not only furnishes Food* for living creatures, but it also extracts from the air that which would be destructive to animal life. It, therefore, is the means of preserving the atmosphere in a pure state for the well-being of the earth's inhabitants.

20. *When the Water which is within a Plant becomes Frozen*, the plant withers, because the water ceases to circulate.

21. *As Snow usually falls before Severe Frost begins*, it keeps the heat of the ground from passing out into the air, and protects the roots of plants and grasses; hence the farmer always welcomes a heavy fall of snow; for the wheat sown in the autumn is protected and nourished by the snowy covering.

"He sendeth forth His commandment upon earth; His word runneth very swiftly. He giveth snow like wool."

22. The *Soil contains Ingredients* necessary to the life of every plant, whether it be the shade or fruit tree, the cotton or tobacco plant, corn, sugar-cane, or potato; and, as the animal body is so constituted as to draw from its food all the elements necessary to the growth of bone and flesh, so the plant draws from water, air, and soil, the different substances required for the growth of wood, leaves, bark, flowers, and fruit.

23. *Besides Soil, Moisture, and Heat*, plants require the light of the sun.

24. The *Light of the Sun assists* in preparing their nourishment, gives them their green color, and causes their leaves and blossoms to open, and their fruit to ripen.

25. The *Grape does not become Fully Ripe in England and Northern France*, because of heavy fogs, which hinder the action of the sun's rays.

26. *All Animals do not eat the same kind of Food*, neither do different plants and trees draw from the soil exactly the same substances.

27. *Each Variety of Plants must be supplied* with the food or elements, adapted to its wants, or it will not flourish.

28. *This is why the Farmer does not sow the same Seed* in the same field every year, and why he manures the soil; for, otherwise, it would soon become exhausted of the elements required specially by the plant which springs from that seed.

29. *Plants thrive* only where the soil allows the roots to spread, and the air and water to penetrate to them; therefore they do not flourish on rock, or in hard, compact clay.

30. *When the Farmer fails to respond to these Laws*, he is soon reminded of his neglect by the appearance of weeds, which seem to call upon him to uproot them; this done, the soil is loosened, and the labor of the industrious husbandman is recompensed by an abundant harvest.

31. *Plants are greatly dependent* upon the moisture and gases contained in the atmosphere.

32. *Some Plants flourish with their Roots either in the Soil*, or in water alone, as the hyacinth. The "air plant" grows without either soil or water, the air affording sufficient nutriment for its growth.

Hyacinth.

Seed of a Maple Tree.

Seed of the Thistle.

33. *Vegetation is extended by the Winds and Water*, which carry seeds to great distances.

34. *For this Purpose* some seeds are provided with a kind of wing, some with a downy substance, and others with a waterproof covering; but the distribution of the useful plants is accomplished chiefly by man.

35. The *Potato was first found* in Peru, and was afterwards taken from Virginia to England by Sir Walter Raleigh, in 1586. It is now cultivated in nearly every part of the world.

36. *Wheat, Rye, and Oats* came, probably, from the western part of Asia.

37. The *Seeds of some West Indian Plants* have been carried by the Gulf Stream to the western and north-western shores of Europe; while, on the other hand, the vegetation of one region may be kept distinct from that of a neighboring region by intervening mountain ranges, or deserts.

38. *Vegetation prevents* the soil from being washed away and injured by the rains.

39. The *Winds not only supply Moisture to the Plants*, but they also remove it when the quantity is superfluous.

40. *Plants are distributed* with reference to climate. In the *Hot Zone* grow rice, sago, bananas, dates, cocoanuts, and yams; in the *Temperate Zone*, wheat, rye, Indian corn, oats, and potatoes; while the *Polar Regions* are almost destitute of food plants.

41. The *Climate of the Torrid Zone* not only affords the most luxuriant vegetation, but keeps the trees and plants in leaf throughout the year; while, in the other zones, vegetation diminishes with the distance from the Equator, and the leaves fall every year, at the approach of winter.

Chart showing that the Luxuriance of Vegetation diminishes toward the Poles. Trees and Productions of the Zones.

42. *In the Torrid Zone*, are the gigantic banyan tree, which covers more than seven acres, and the lofty palm, reaching the height of two hundred feet; while in the Frigid Zone, there are found only dwarfed trees, low plants, and mosses.

43. *Vegetation in the Northern Hemisphere* extends further north on the western sides of the continents than on the eastern, owing to the agency of the south-west winds which blow over the warm currents of the ocean.

44. The *Forest Trees of the Temperate Zones* are mostly deciduous—that is, their leaves fall in the autumn; some, however, are evergreen, or indeciduous.

45. The *Productions peculiar to the Temperate and Frigid Zones* do not generally thrive in the hot zone, even if transplanted there, unless they are placed in elevated situations, where the climate corresponds with that of higher latitudes.

46. *Apples, Pears, and Grapes* belong to the Temperate Zone, and thrive in the Hot Zone only at an elevation of from 600 to 1,000 feet.

47. The *Productions of One Zone* are not separated from those of the adjoining zone by any distinct line, the change from one zone to another being gradual.

48. *From the Base to the Summit of a Lofty Mountain*, vegetation varies with the elevation; on its sides are the same gradations of climate, with their characteristic varieties of plants and trees, that exist on the earth's surface between the latitude of the mountain and the Poles.

49. The *Mountains and Valleys in the State of California* afford every variety of climate, with fruits peculiar to every zone. There flourish the olive, the fig, the date, the grape, the pine-apple, the peach, the apple, and the pear; besides all varieties of grain. In the forests grow mammoth trees, many being from 300 to 400 feet high, and from 25 to 35 feet in diameter.

50. Of what use is vegetation? What kinds of food are adapted to the inhabitants of the Temperate Zones? The Frigid? The Torrid Zone? In what zone do bananas, cocoa nuts, and dates grow? What zone is most favorable to grain, apples, and grapes?

Do different kinds of plants receive their nourishment from the same ingredients of the soil?

In what zone is vegetation the most abundant?

Mention some of the uses of trees and plants. Of snow.

What two elements are necessary to the growth of all plants?

Name the principal trees and plants of the Torrid Zones. Of the Temperate Zones. Of the Frigid Zones. (See illustration on first column.)

What effect have the winds upon the climate and productions of California?

What can you say of the trees of California?

On which side of North America does vegetation extend further north?

In what part of the Torrid Zone could you find the climate and productions of the Temperate Zone?

51. *The land which forms the continents* was, at first, but slightly elevated above the surface of the water, and became covered with plants and heavy trees, such as are shown in the illustration on the following page—second column.

52. *Portions of the Earth's Surface* would sink below the water, and their masses of vegetation, which were covered with sand, clay, etc., now appear in the form of coal.

53. *Such was the Formation of the Important Coal Fields* of the world.

54. *In North America*, one extends from Pennsylvania to Alabama, the workable area of which is estimated at 60,000 square miles; another large field extends from Illinois to Texas. Coal abounds also in New Brunswick, Nova Scotia, Prince Edward's Island, and Newfoundland.

55. *In the Old World*, vast beds exist in Great Britain, France, Belgium, Spain, Germany, Hungary, and China.

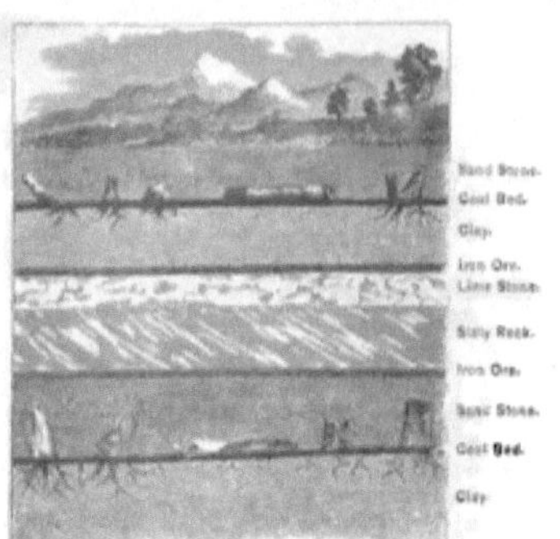

Interior or Sectional View of the Coal Regions of Pennsylvania, showing Strata, which resulted from Successive Submergences of the Surface. The Trees whose Stumps are here represented, flourished at the Earth's Surface in Periods long past.

56. *By Digging downward* in the coal regions, various strata are met with, as shown above; they do not consist of the same materials, nor do they lie in the same order, in all places.

57. *The Distribution of Coal* in various parts of the earth, plainly indicates that its importance to man was anticipated by the Creator. Even the necessity for coal, in the working of iron ore, was provided for by Him; this is observed in the remarkable association of the two.

58. *The Dirt-beds which contain the Roots of Trees and Plants*, formed, at some period, the surface soil which supported vegetation; and the greater the vegetable mass that was submerged, the thicker would be the coal bed; and, while a coal bed extends over considerable space, it is generally much thinner than the strata of sand, clay, and stone, which may be above or below it.

59. *Many Stumps of Large Dimensions*, and with very extended roots, have been found both in America and England, transformed into coal; the stumps retaining their shape and the natural roughness of the bark.

60. The *Vegetation of which Coal was formed*, included the trees and plants of the forests and marshes.

61. *Vegetation which undergoes Decay* on the surface of the earth serves to enrich the soil.

62. *Vegetation which entered into the Formation of Coal* must have been entirely submerged through long periods of time.

63. *Had there been no Submergence* of vegetation, we would not now be provided with coal.

64. The *Different Coal Beds*, lying one below the other, show how often that part of the surface was above the water level, and covered with vegetation.

65. *In Nova Scotia*, there have been discovered nineteen parallel seams of coal, varying in thickness from two inches to four feet.

66. *At the present Rate of Consumption of Coal*, it is estimated that the coal fields of Pennsylvania alone, could meet the demand of the whole world for more than 1,000 years.

Appearance of Parts of the Earth's Surface at the Commencement of the Age of Reptiles. The Fern with other Trees and Plants here represented entered largely into the Formation of Coal.

Section XIX.

ANIMALS; THEIR CREATION AND USES.

1. *Vegetable and Animal Life existed* long before the creation of man, and mutually contributed to each other's support and nourishment; vegetation sustaining animal life, and the decay of animal bodies and substances, through long ages, adding to the fertility of the soil.

2. *Soil that is destitute of Decomposed Animal or Vegetable Substances* is very poor, and will yield little or no vegetation; such was the condition of vegetable life at its commencement; such, also, was the beginning of animal life—very inferior in character and form.

3. *An Improvement in the Quality of the Soil*, caused an improvement also in the varieties of plants; following which came different and improved species of animals.

4. *Geologists show that the Animals which were first created* were very different from those we now see upon the land.

5. *Those first formed* were of the simplest construction, hardly distinguishable from plants.

6. *Different Kinds or Classes of Animals* followed each other; each class being superior in construction, powers, and usefulness, to those which preceded it.

7. *Throughout the Works of Nature*, we see the leading law of development—improvement by successive steps.

8. *According to this Law*, from a small seed springs a tender plant, which enlarges *gradually* until it becomes a great tree.

9. The *Mighty River* started upon its course as a mere rivulet, which was formed from a trickling spring.

RADIATES.

Jellyfish. Starfish. Actinia. Coral. Medusa. Polyps. Actinia.

10. *Animal Life first appeared in the* form of *Radiates.* After them came *Mollusks,* then *Articulates;* after these there followed in order, *Fishes, Reptiles,* and *Mammals.* Last of all came *Man.*

11. *A Knowledge of the Animals which preceded Man* is obtained by digging into **stratified** rock, where their forms, sizes, and construction **are distinctly** observed. (*See page* 8, *paragraph* 10.)

12. *Radiates,* in construction, resemble a flower or plant, but differ from them in having a mouth and stomach. Their bodies are nearly transparent, and seem only to float or rest in water.

MOLLUSKS.

Nautilus. Squid. Scallop. Clam. Oyster. Snails.

13. *Mollusks* are those which have soft bodies without bones or skeletons; some are naked, while others are enclosed in shells for their protection. Of the latter, oysters, clams, and snails furnish examples.

14. *Articulates* are characterized by jointed or articulated coverings consisting of a series of rings: they comprise such animals as worms, crabs, lobsters, spiders, and winged insects.

ARTICULATES.

Common House-fly. Beetle. Mosquito. Caterpillar. Butterfly. Grasshopper. Lobster.

15. *Following the Creation of Articulates was that of Vertebrates,* which embrace all animals having a backbone.

16. The *First Vertebrates* were fishes, then reptiles, birds, and mammals.

17. *Mammals* are those animals which breathe **with lungs,** suckle their young, and have warm blood. They **include** Mankind (*bimana—having two hands*), the Monkey (*quadrumana—having four hands*), and the following named animals

CARNIVORA, OR FLESH EATERS.			RUMINANTS, OR CUD-CHEWERS.		RODENTS, OR GNAWERS.	
Lion,	Panther,	Bear,	Ox,	Deer,	Hare,	Beaver,
Tiger,	Dog,	Walrus,	Sheep,	Camel,	Rabbit,	Rat,
Leopard,	Cat,	Seal.	Goat,	Giraffe.	Squirrel,	Mouse.

PACHYDERMS, OR THICK-SKINNED ANIMALS.		EDENTATES, OR TOOTHLESS.	CETACEA, OR SEA MAMMALS.	INSECTIVORA, OR INSECT EATERS.
Elephant,	Horse,	Sloth,	Whale,	Mole,
Hippopotamus,	Zebra,	Ant-eater,	Porpoise,	Bat,
Rhinoceros,	Hog.	Armadillo.	Dolphin.	Hedgehog.

18. *Animals of the Different Zones.*

IN THE ARCTIC REGIONS OF BOTH HEMISPHERES.

The Reindeer, Polar Bear, Whale, Seal.

IN THE TEMPERATE ZONES OF BOTH HEMISPHERES.

Horse, **Ox, Sheep, Deer,** Wolf.

IN THE TEMPERATE ZONE.

North America,	Grizzly Bear,	**Bison,**	**Puma,**		
Europe,	Brown Bear,	**Chamois,**	**Wild Boar,**	Stag.	
Asia,	Tiger,	**Camel,**	Musk,	Deer	Sable.

IN THE TORRID ZONE.

South America,	Jaguar or American Panther,		Puma,	Tapir,
	Llama,	Alpaca,	Sloth,	Monkey.
Asia,	**Camel,**	Tiger,	Elephant,	Rhinoceros,
	Asiatic Lion,	Panther	Crocodile,	Monkey.
Africa,	African Lion,	Camel,	Hippopotamus,	Antelope,
	Camelopard or Giraffe,		Zebra,	Hyena,
	Leopard,	Orang Outang,	Ape,	Monkey.

19. It is believed that the *Submergence, at Different Periods,* of vegetation which entered into the coal formations, occurred before the creation of birds; and with vegetation, sank also vast collections of animal bodies, such as mollusks, insects, fishes, and reptiles, which contributed largely to the formation of the strata beneath the present surface of the earth.

20. The *Earth yields Productions and Species of Animals* peculiar to each region or climate.

21. The *Largest Animals* are in the hot regions; they are the elephant and hippopotamus, whose covering is a tough skin, almost entirely destitute of hair; while, in the Arctic regions, where it is too cold for the horse and the ox, live the reindeer and Polar bear, thickly covered with hair, to protect them from the severe cold.

22. The *Near Approach of America to Asia,* at Behring Strait, has given to the Arctic regions of both continents the same species of animals.

23. The *Reindeer and Polar Bear* abound in the Arctic regions of North America, Europe, and Asia.

24. *Animals are adapted* to the zones and districts which they inhabit; their wants and uses are wonderfully fitted to the circumstances in which they are placed.

25. *In the Temperate and Warm Zones* is found the horse, which is the most useful of all animals.

Laplanders on their Sleds drawn by Reindeers.

26. *In the Frozen Regions of the North*, are found the reindeer and the seal.

27. The *Reindeer* constitutes almost the entire wealth of the Laplander, furnishing him with flesh and milk for food, and drawing his sledge over vast fields of snow.

28. *These Animals obtain their Food* from mosses and low plants, for which they root through the snow, like swine in a pasture.

29. The *Esquimaux derive their Support* from the seal, and exert their greatest energies in the capture of this aquatic mammal.

30. The *Flesh and Fat of the Seal* are used for food; its oil, for light and fuel; the skins are made into clothing, leather, boats, and tents, and form an important article in the fur trade.

31. *Seals are found* in large numbers on fields of floating ice near the coast of Greenland.

32. The *Camel* was made for the desert, where the burning climate and the absence of water render all other animals useless to man.

33. *Providence has given to the Camel* a kind of reservoir or system of cells in which to carry a supply of water sufficient for a long journey; it is also furnished with sharp teeth to cut the few tough shrubs of those barren tracts; and, that it may not be suffocated by the driving sand and dust, its nostrils are so formed as to allow respiration without admitting sand. Its feet are provided with a kind of pad or cushion which prevents their sinking into the soft and yielding sand.

34. *Some Animals inhabit* the dry land, some the water, some fly in the air, and others have the power of living either on land or in water. These last are called amphibious.

35. *A Bird was not formed to live in Water*, like a fish, hence it is not covered with scales; a fish cannot live in the air and find its food among the trees; therefore, it is not provided with feathers and wings; the elephant, the horse, and the ox are unlike both the bird and the fish; but according to their several requirements and uses, they have received their forms, powers, and places.

36. *Animals, like Plants, abound* most in the hot zone, and least in the frigid.

37. The *Surpassing Abundance*, in South America, of vegetation and of the lower species of animals, such as insects and reptiles, is attributable to the excessive heat and moisture of its tropical regions.

Section XX.

MANKIND; THE RACES.

1. "Thus the heavens and the earth were finished, and all the host of them. And the Lord God formed man of the dust of the ground, and breathed into his nostrils the breath of life; and man became a living soul."

2. *For what Purpose* was man created? (*See page* 5.) Was man created before, or after, animals? Why? Were grass, plants, and trees made before, or after, the creation of animals? Why?

3. *Man is distinguished from all other Animals*, not by his form only, but by his powers of reason and speech. He acknowledges the infinite goodness, wisdom, and power of the Creator, and seeks to advance continually in wisdom and happiness.

4. *Man's Constitution* is such that he is capable of living in any latitude, from the hot to the frozen zone; or at any elevation between the level of the sea and the region of perpetual snow on the sides of mountains.

5. *However Extreme may be the Coldness* of the climate which man enters, his dominion over the animal, vegetable, and mineral kingdoms enables him to procure from them clothing and fuel, which compensate for the lack of solar heat.

6. *While mere Animals are restricted to a Few Varieties of Food*, man partakes of the fruit and vegetables of the soil, and of the flesh of creatures which inhabit the land, the water, and the air.

7. *Mankind is divided into Five General Classes*, or races: the Caucasian, or white race; the Mongolian, or yellow race; the Ethiopian, or black race; the Malay, or brown race; and the American Indian, or red race.

8. The *Races are distinguished from each other* by the color of the skin, kind of hair, and structure of the body and the skull.

9. *These Differences* are produced chiefly by differences in climate, food, and pursuits.

10. The *Influences of these Conditions* upon the physical and mental characteristics of man are vast and unavoidable.

11. *Change the Climate of a Country* either in degree of temperature or of moisture, and a change will be effected also in the character of its vegetation, in the number and kinds of its animals, and in the temperament and pursuits of the inhabitants.

12. The *Condition of a Nation* would be affected by a material change in its systems of rivers, canals, and railroads.

13. *Improved Means of Intercommunication* serve to advance the civilization, education, and prosperity of the people, and to promote a spirit of national unity.

14. *This is obvious in the United States*, where constantly increasing lines of travel by railroads, steamboats, and canals, together with elaborate postal and telegraph systems, contribute largely to the growing power of this republic.

15. The *Depressing Effects of the Absence of these Means* of development are observed in the condition of Africa and the greater part of Asia.

16. *Races and Nations are adapted to the Climate* of whatever portion of the earth they inhabit.

17. The *Hindoo and the Ethiopian* prefer their hot zone, with its light, vegetable food.

18. The *Esquimaux and the Laplanders* cling with strong attachment to their boundless fields of snow, obtaining their subsistence from the animals and fish of the Arctic regions.

19. The *Greenlanders* have their habitation between 70° and 80° north latitude, while the Red Men of South America, and the Blacks of Africa, live under the burning sun of the Equatorial regions.

20. The *White Inhabitants of North America and Europe*, accustomed to a temperate climate, can live in either of those extremes, and on almost every variety of food.

21. *Europe Colonized the Temperate Zone* of North America with wonderful success, but the results of her efforts in other zones have been, comparatively, failures.

22. *In the Tropical Part of Asia*, is British India, which is celebrated for the richness of its productions,—the cotton-plant, sugar-cane, silk, and all varieties of fruits, besides gold, diamonds, precious stones, and nearly all the metallic ores; but, notwithstanding England's influence and authority in that section for more than a century, there is yet only one white inhabitant for every 3,000 natives.

23. *In the Tropical Regions*, the inhabitants subsist, to a great extent, upon the spontaneous yield of the soil; this, together with the enervating influence of the oppressive heat, causes them to lack energy, industry, and patriotism.

24. *In the Frozen Regions*, the inhabitants are dwarfed both in physical stature and mental powers; this is owing to the severity of the climate, with the absence of natural productions and of inducements to labor.

25. *Hardships, Want, and Continued Cold* in the Frigid Zone, and luxury, indulgence, and continual heat in the Torrid, retard the development of their inhabitants.

26. *Both of these Regions* lack that *diversity* of climate and of other conditions, which is necessary to the promotion of individual and national prosperity.

27. *In the Temperate Zones* are enjoyed the greatest blessings which the earth affords. Their lands are neither parched nor icebound; neither teeming with enervating luxury nor stinted to shrubs and mosses; their position on the globe, their systems of mountain ranges, ocean currents, and their change of seasons, combine to promote among the people, that spirit of energy and enterprise essential to their development and happiness.

28. *It is in the Temperate Zone* that the climate and soil both demand and reward the exercise of man's energies, making vast plains to become fields of smiling plenty and drawing from rugged mountains incalculable riches.

29. *Vegetable and Animal Nature* increases in luxuriance and strength with distance from the Poles, but the distribution of the human races is different, in this respect.

30. *Man has attained the Highest State of Development*, physically, mentally, and morally, in the North Temperate Zone, or between the parallels of 30° and 60° north latitude.

31. *Within these Lines*, are the United States of America and all the leading nations of Europe and Asia.

Characteristics of the Torrid, North Temperate, and North Frigid Zones.

32. *In the Temperate Zone of Asia*, the human race had its birth, and here also Christianity was first given to man.

33. The *Temperate Zone does not Encourage Idleness*, and, therein, is unlike the Tropical; but it fully rewards labor, industry, and skill, and in that respect it differs from the Frigid.

34. The *Caucasian*, or white race, comprise the most powerful and enlightened nations of the world.

35. *They inhabit* nearly all that part of North America which lies south of the parallel of 50° north latitude, or that part south of the northern boundary of Canada; along the coasts of South America; the greater part of Europe; western and south-western Asia; northern and north-eastern Africa.

36. The *Mongolians*, or yellow race, have thin, coarse, and straight hair, low foreheads, wide and small noses, and thick lips.

37. *They are more numerous* than any other race.

38. The *Mongolians inhabit* the Arctic regions of both continents, and all Asia, except its western and south-western parts.

39. The *Chinese, Japanese, and Esquimaux* belong to the yellow race.

40. The *Ethiopians*, or black race, thrive in the heat and dampness of the tropics, where the white man soon dies.

41. *They inhabit* nearly all that part of Africa which lies south of the Great Desert.

42. The *Egyptians, Abyssinians, and Berbers*—the inhabitants of Barbary—are Africans, but not Negroes. They belong to the Caucasian race.

43. The *Malays* are of a reddish brown color; their hair is black, straight, coarse, and abundant.

44. The *Malays* are treacherous, ferocious, and less sensible to pain than the other races.

45. *They inhabit* the Malay Peninsula, Sumatra, Java, New Zealand, and many other islands of the Indian and Pacific Oceans.

46. The *American Indians*, so called by Columbus, are copper-colored, tall in stature, and have straight, black hair.

47. Before the arrival in America of the whites, the Western Continent was inhabited by the red men, excepting, however, in the Arctic regions and Greenland, which are inhabited by the Esquimaux.

48. The *Esquimaux* are classed among the Mongolians, in which race many authorities include also the Indians of America.

49. The *American Indians*, in disposition, are melancholy, revengeful, and jealous, and feel bodily pain less acutely than the whites.

50. The *Red Men and the Esquimaux* of America entered that division from Asia, probably in the direction of Behring Strait.

51. The *Human Family had its Origin* in Western Asia, whence it extended into all lands. From the race that moved westward and peopled the lands bordering on the Mediterranean Sea, sprung nations celebrated in ancient history for their progress in civilization and learning.

52. *In Africa*, were ancient Egypt and Carthage; and in Europe, were Greece and the Roman Empire.

53. The *Wave of Progress and Power* continued to roll westward to the Temperate regions of the New World, now the United States of America.

"WESTWARD THE COURSE OF EMPIRE TAKES ITS WAY."

54. *Columbus sailed Westward*; and, by his discovery of the Western Continent, two worlds became acquainted with each other, for their mutual development and advantage. One contributed its vast natural resources; the other, its blessings of civilization and vigor of intellect.

55. The *New World was near enough to the Old* to receive aid while in its infancy, and far enough from it to demand of its new inhabitants the most active employment of their energy and skill toward the development of its resources.

56. The *New World has grown* in usefulness, greatness, and influence with wonderful rapidity.

57. The *North Temperate Zone of America* is vast in vegetable, mineral, and commercial wealth, and contains a people renowned for their energy, enterprise, and achievements, both in peace and in war.

58. *As each Successive Period in the Creation of the Earth* was marked by improvement, so the American Nation is recognized as rising above all others in the sphere of usefulness, development, and influence.

59. The *Productive Plains of the Center and South*, the manufacturing region of the north-east, the broad plains and rich mines of the west, united by easy lines of communication and occupying positions perfectly adapted to each other—plainly show that Providence designed this nation to be ONE AND INDIVISIBLE.

NOTE.—The teacher will here turn to the "INDEX TO CONTENTS ARRANGED AS A GENERAL REVIEW OF PHYSICAL GEOGRAPHY," which may be found near the end of the book, and divide it into lessons of convenient length for the class.

PART

GEOGRAPHY.

DEFINITIONS.

1. *GEOGRAPHY* is a description of the earth's surface. Physical Geography describes the natural features—the land, water, currents, soil, climates, and their effects upon the inhabitants. Civil, Political, or Descriptive Geography treats of the artificial divisions—republics, empires, kingdoms, states, etc.

NATURAL DIVISIONS.

2. *Water Covers* three-fourths of the earth's surface; land, one-fourth.

3. *A Continent* is the largest natural division of the land. There are two continents—the eastern, composed of Europe, Asia, and Africa; and the western, composed of North and South America. The term continent is frequently applied to Australia and to each of the divisions just mentioned.

4. *An Island or Isle* is a portion of land entirely surrounded by water. Several islands together are called a group; in a line, a chain. An islet is a small island. A ledge of rocks at or near the surface of the water is called a reef or keys.

5. *A Peninsula* is a portion of land almost surrounded by water.

6. *An Isthmus* is a narrow neck of land joining two larger portions of land.

7. *A Cape or Headland* is a point of land extending into the water. A high and rocky cape is called a promontory.

8. *A Mountain* is a vast elevation of land; a hill is a small elevation. A chain or range is a long elevated ridge, or several mountains extending in a line.

9. *A Mountain System* is a number of chains grouped together.

10. *A Peak* is a single mountain, whose top appears pointed.

11. *A Volcano* is a mountain or opening in the earth's crust through which issue fire, smoke, ashes, lava, steam, &c.; the opening is called a crater.

12. *Mountain Passes* are the lowest parts of a chain, where travelers can cross.

13. *A Valley* is the land between hills or mountains, or at their base. A vale is a small valley.

14. *A Plain or Lowland* is a level tract of land. The grassy, treeless plains of North America are called prairies or savannas; of South America, llanos (*lyah'noce*) and pampas; of Russia, steppes (*steps*). The forest plains of the Amazon are called silvas. A swamp, marsh, moor or fen is a tract of land usually or occasionally covered with water.

15. *A Plateau or Table Land* is a plain at a considerable elevation above the level of the sea.

16. *A Desert* is a barren region of country; the fertile spots are called oases.

17. *An Avalanche* is a large mass of snow, ice, and earth, sliding or rolling down a mountain. When the mass consists of earth alone, it is called a landslide.

18. *A Glacier* (*gla'seer*) is an immense mass of ice and snow formed in the region of perpetual snow, and moving slowly down the mountain slope or valley, bearing with them gravel, sand, and masses of rock.

19. *A River Basin* is the portion of land which is drained by a river and its tributaries.

20. *A Water Shed* is the mountain chain or ridge of land which separates one basin from another, and from which the rivers flow.

21. *A Delta* is the lowland between the several mouths of a river. It is composed of the soil which has been carried down by the stream, and deposited at its mouth (see page 30, paragraph 44).

22. *A Coast or Shore* is the edge of land adjacent to the water.

23. *An Ocean* is the largest natural division of the water.

24. *A Sea* is the division next in size to an ocean. A sea containing many islands is called an archipelago (*ark*). The Sargasso Sea is that part of the Atlantic Ocean between Africa and the West Indies, which contains great quantities of seaweed.

25. *A Gulf or Bay* is a body of water extending into the land. Harbors, havens, ports, roads, and roadsteads are places where ships may anchor safely.

26. *A Strait* is a passage connecting two larger bodies of water. A channel is a broad strait.

27. *A Sound* is a shallow channel or bay.

28. *A Lake* is a body of water almost surrounded by land.

29. *A River* is a stream of water flowing through the land; its head or source is its beginning, and its mouth is its end, or where it flows into another body of water. The right bank of a river is on your right side as you descend the river; its left bank is on the left side. *Up* a river is toward its source; *down* a river toward its mouth.

30. *An Estuary, Firth or Frith* is a narrow and deep inlet of the sea, at the mouth of a river.

31. *A Cañon* (*pronounced and also spelled Canyon*) is a gorge or ravine between high and steep banks, worn by a stream.

32. *A Confluence* is the junction of two or more rivers.

33. *A Cataract* is a large body of water falling over a precipice; a cascade or waterfall is smaller than a cataract.

34. *A Rapid* is the descent of a stream over an inclined part of its bed.

35. *A Loch* in Scotland, or a lough in Ireland, is a lake or bay.

CIVIL OR POLITICAL DIVISIONS.

36. *A Republic* is a country whose laws are made and executed by men elected, from time to time, by the people. We live in a Republic—the United States. The legislative or law-making body is Congress, which is composed of the Senate and House of Representatives; the executive power is vested in the President.

37. *An Empire* is a country governed by an emperor, or an extensive region comprising several countries, under one ruler.

38. *A Kingdom* is a country governed by a king.

39. *A Monarchy* is a government in which the supreme power belongs to one person, called a monarch. Emperors and kings are monarchs, and their governments monarchies.

40. *A Limited or Constitutional Monarchy* is a government in which the power of the ruler is limited by laws.

41. *An Absolute Monarchy or Despotism* is a government in which the power of the ruler is unlimited.

42. *The Divisions of a Republic* are usually called States.

43. *The Government of a State* resembles that of a Republic; its laws being made by the legislature, and executed by the Governor.

44. *A Territory of the United States* is a portion of the country not organized as a State; its governor and legislature are appointed by the President and Senate of the United States.

45. *The Divisions of Countries* are called States, Provinces, Departments, Counties or Shires.

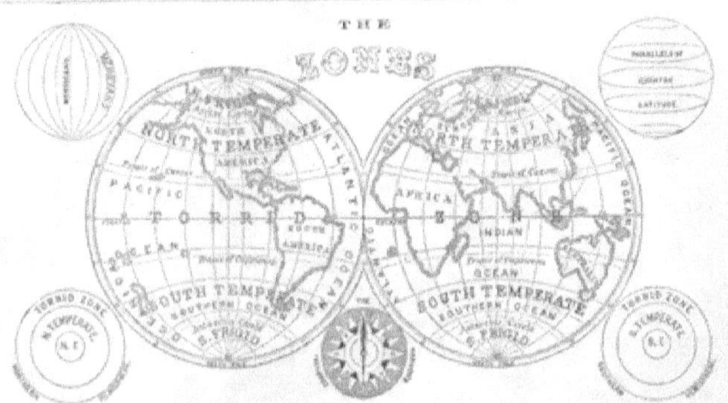

CIRCLES, ZONES, LATITUDE, ETC.

1. ***An Angle*** is the opening between two lines that meet. There re three kinds: right, acute, and obtuse.

2. ***A Right Angle*** is formed by one line meeting another perendicularly.

3. ***An Acute Angle*** is less than a right angle.

4. ***An Obtuse Angle*** is greater than a right angle.

5. ***Parallel Lines*** are those which extend in the same direction ithout approaching each other.

6. ***A Horizontal Line*** is one that is parallel with the horizon, r with the surface of water at rest.

7. ***A Triangle*** is a figure which has three sides and three ngles.

8. ***A Circle*** is a figure enclosed by a curve line, every part of hich is equally distant from the centre; the curve line is called e circumference, or ring, and also circle.

9. ***The Diameter*** of a circle is a line drawn through the centre om one side to the other.

10. ***The Radius*** is a line drawn from the centre to the circumference; two radii equal the diameter.

11. ***An Arc*** is any portion of the circumference of a circle.

12. ***A Chord*** is a straight line drawn from one end of an arc to e other.

13. ***A Quadrant*** is a quarter of a circle; a semi-circle is half a ircle.

14. ***A Circle*** is, or is supposed to be, divided into 360 equal arts; these parts are called degrees.

15. ***A Degree*** is a three hundred and sixtieth part of a circle. he length of a degree varies according to the size of the circle.

16. ***A Degree of the Largest Circle*** which can be drawn round the earth, is about 69½ miles in length.

17. ***A Degree, marked (°), is Divided*** into 60 equal parts, alled minutes ('), and each minute into 60 equal parts, called econds (").

18. ***A Semi-Circle Contains*** 180°, and a quadrant 90°.

19. ***A Great Circle*** divides the earth into two equal parts. hat which divides the earth into northern and southern hemipheres is called the equator.

20. ***A Small Circle*** divides the earth into two unequal parts. ll circles drawn parallel with the equator are small circles, called arallels of latitude.

21. ***The most Important Small Circles*** are the Arctic and Antarctic circles, the Tropic of Cancer, and the Tropic of Capricorn. The Arctic and Antarctic Circles are called Polar Circles.

22. ***A Globe or Sphere*** is a round body, whose surface, in every part, is equally distant from the centre.

23. ***The Axis*** of the earth is the line or diameter on which it revolves. The two points where the axis meets the surface are called the poles; that in the centre of the northern hemisphere is called the North Pole; that in the centre of the southern hemisphere, the South Pole.

24. ***Semi-Circles*** drawn on the surface from the North to the South Pole are called Meridians.

25. ***The Distance*** of any place from the equator is its Latitude.

26. ***The Latitude*** of all places on the equator is 0°, and of each pole 90°.

27. ***Distance East or West*** of an established meridian is called Longitude; the degrees of which are marked on the equator or on the upper and lower sides of a map. Longitude is usually reckoned from the Meridian of Greenwich, near London, and from the Meridian of Washington.

28. ***The Greatest Longitude*** a place can have is 180°.

29. ***Zones*** are five regions or belts into which the earth's surface is divided by the two Tropics and the two Polar Circles. *Name them.*

30. ***The Hottest*** is the Torrid Zone, and the coldest are the Frigid Zones.

31. ***To every place Within the Torrid Zone*** the sun is vertical at certain times during the year.

32. ***The Tropics Mark*** the limit beyond which no place can have a vertical sun. They are 23½° from the equator.

33. ***The Polar Circles*** mark the limit beyond which the day or the night may be more than 24 hours long. They are 23½° from the Poles.

34. ***The Sun is Visible*** at one pole, and invisible at the other pole, during six months of the year.

35. ***Days and Nights*** are equal—12 hours each—throughout the earth on the 23d days of March and September; they are always equal to places on the equator.

36. ***In what Zones is*** North America? South America? Europe? Asia? Africa? Australia?

(*For Astronomical Geography, see page 108.*)

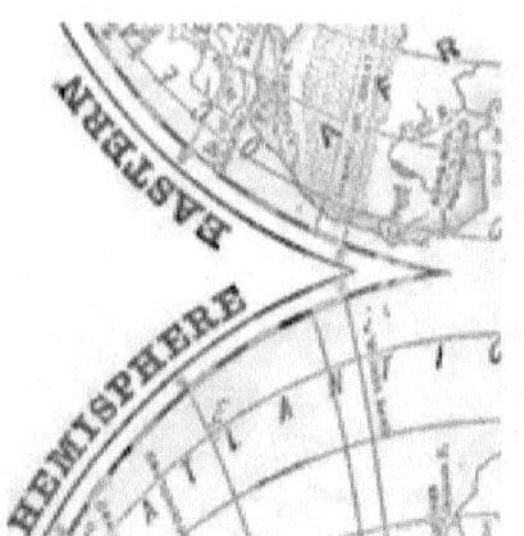
EASTERN
HEMISPHERE

EXERCISES ON THE MAP.

How many continents are there? Name them.
What are the grand divisions of the Western Continent?
What are the grand divisions of the Eastern Continent?
Which is the largest of the divisions? The smallest?
What division extends farthest north? Farthest south?
Is one hemisphere larger than the other?
Is one continent larger than the other? Which is the larger?
In what direction does the Western Continent extend? The Eastern?
In what direction does each division extend?
What divisions are entirely north of the Equator?
What divisions are partly south of the Equator?
Which has its greatest part south of the Equator?
Is the greater part of Africa north or south of the Equator?
What part of North America is the narrowest? Of South America? Of Africa? Greenland? Hindoostan?
What divisions have their greater parts in the Torrid Zone?
What divisions are chiefly in the North Temperate Zone?
What lands are in the South Temperate Zone?
What division is almost entirely in the North Temperate Zone?
In what zones is North America? South America? Asia? Australia?
What three peninsulas in the southern part of Asia?
In what direction do nearly all peninsulas point?
Name the oceans. Which is the largest?
Where is the narrowest part of the Pacific Ocean?
With what ocean in the north is the Pacific connected?
What connects the Arctic with the Pacific Ocean?
With what ocean in the south is the Pacific connected?
Where does the Atlantic meet the Pacific?
In what part of the Western Continent is the nearest approach of the Atlantic to the Pacific?
Sailing around Cape Good Hope in an easterly direction, from what ocean do you sail, and what ocean do you enter?
What large arms has the Indian Ocean? The Atlantic?
On which side of America is its great mountain chain?
With what coast is it parallel? Name the mountains of Africa. Of Asia.
What ocean receives most of the rivers of the Western Continent?
Mention the principal rivers in the Western Hemisphere. The Eastern.
Mention the largest island in the world. The largest seas, gulfs, and bays.
What general name is given to the islands in the Pacific Ocean? What islands are crossed by the Equator?
What islands off the east coast of Asia? Of America?
What islands on the eastern side of the Atlantic Ocean?
Name the principal deserts. Where are they?
What lands at the center of the Land Hemisphere? What islands at or near the center of the Water Hemisphere?

Bound **NORTH AMERICA.** What two large bays in the north-east?
With what two oceans are these bays connected?
What large gulf south-east of North America?
What large rivers between the Rocky Mts. and Atlantic Ocean? What one between them and the Pacific Ocean?
What is the largest city in North America?
Bound **SOUTH AMERICA.** What joins North and South America?
Are these two divisions alike in possessing great inlets from the ocean? Which has the most unbroken coast line?
Name the principal rivers in South America. Into what do they flow?
Why do no large rivers of South America flow into the Pacific Ocean?
What strait south of South America? What islands?
What is the southern cape? Northern? Eastern? Western?
What is the largest city in South America?
Bound **EUROPE.** In what respect does the coast line of Europe differ from those of South America and Africa?
What islands west? North-east?
With what ocean are the large seas of Europe connected?
What connects the Mediterranean Sea with the Atlantic Ocean?
What is the northern cape of Europe?
Bound **ASIA.** In what directions do the rivers of Asia flow?
What parts of Asia are destitute of rivers?
What is the character of the soil in these regions?
What and where is the largest desert in Asia?
What two seas in the west have no outlets? Are their waters fresh or salt? Why? (See page 31.)
Bound **AFRICA.**
In what part of Africa are its mountains? Rivers?
What great region in the north destitute of rivers?
What cape on the eastern coast? Southern? Western?

REVIEW.

GRAND DIVISIONS.

Where are they? By what waters are they surrounded?

NORTH AMERICA?	EUROPE?	AFRICA?
SOUTH AMERICA?	ASIA?	OCEANICA?

OCEANS.

Where are they? By what lands are they embraced?

PACIFIC? ATLANTIC? INDIAN? ARCTIC? ANTARCTIC?

SEAS.

Where are they? Into what waters do they open?

MEDITERRANEAN?	ARABIAN?	CHINA?	BLACK?
CARIBBE'AN?	CASPIAN?	NORTH?	RED?

GULFS AND BAYS.

Where are they? Into what waters do they open?

MEXICO? GUINEA? HUDSON? BAFFIN? BENGAL?

STRAITS AND CHANNELS.

Between what lands? What waters do they connect?

MOZAMBIQUE? MAGELLAN? GIBRALTAR? COOK'S? BASS?

RIVERS.

In what part of what division do they rise? Into what waters do they flow?

AMAZON? 40*	CAMBODIA? 20	NIGER? 21
MISSISSIPPI? 32	MISSOURI? 31	LA PLATA? 1
YANG-TZE-KIANG? 32	HOANG HO? 30	NILE? 40
COLUMBIA? 12	ORINOCO?	AMOOR?
YENISEI? 30	PARANA? 20	OBI? 20
(yen-e-sey'e.)	(pah-rah-nah'.)	(o'bee.)

MOUNTAINS.

Where are they? In what direction do they extend?

ROCKY? 15*	ANDES?	URAL? 5
HIMALAYA? 29	ALTAI? 11	ATLAS? 15
(him-a-lā'a.)	(al-tī'.)	KONG? 3

ISLANDS.

Where are they? By what waters are they surrounded?

NEW'FOUNDLAND?	GREENLAND?	WEST INDIES?
MADAGASCAR?	JAPAN IS.?	SOCIETY IS.?
TERRA DEL FUEGO?	CAPE VERD IS.?	NOVA ZEMBLA?
BRITISH IS.?	FRIENDLY IS.?	ICELAND?
SANDWICH IS.?	NEW ZEALAND?	JAVA?
TASMANIA?†	NEW GUINEA?	BORNEO?
PHILIPPINE IS.?	AUSTRALIA?	SUMATRA?
(fil'ip-in.)	(aws-tray'li-a.)	(su-mah'trah.)

CAPES.

From what part of what land do they project, and into what waters?

FAREWELL?	BLANCO?	VERD?	HORN?
GOOD HOPE?	ST. LUCAS?	NORTH?	COMORIN?
GUARDAFUI?	ST. ROQUE?	GALLINAS?	(kom'o-rin.)
(gwar-dah-fwee'.)	(sent roke'.)	(gal-lē'nas.)	

---

* The numbers in lists of Rivers show the length in *hundreds of miles*; those in lists of Mountains, show the height of highest peaks, in *thousands of feet*: e.g., the length of the Amazon is 4000 miles, the height of the Himalayas is 29,000 feet. † Or Van Diemen's Land.

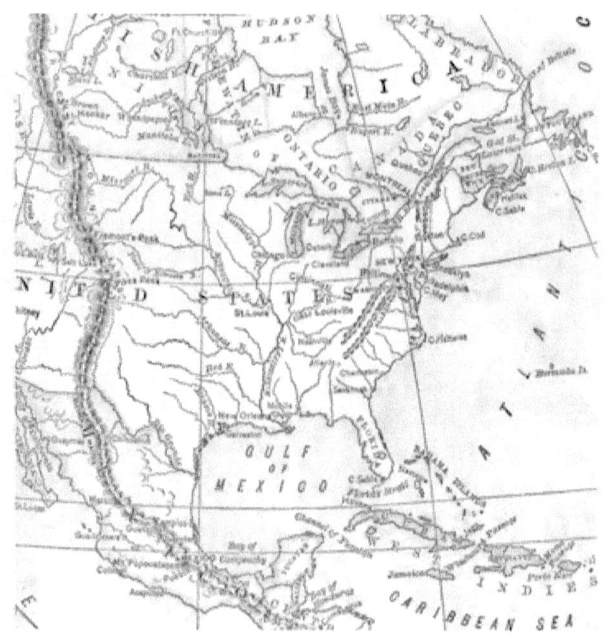
HUDSON BAY
GULF OF MEXICO
CARIBBEAN SEA
LABRADOR
St. Louis
Louisville
New Orleans
C. Hatteras
Bermuda Is.
FLORIDA
C. Sable
Jamaica
Florida Strait

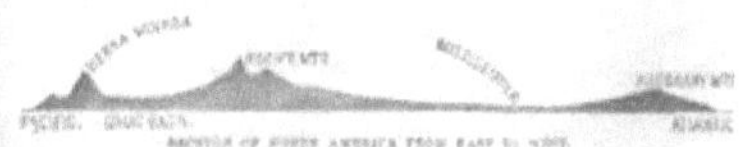

SECTION OF NORTH AMERICA FROM EAST TO WEST.

EXERCISES ON THE MAP.

In what zones is NORTH AMERICA? In which is the greater part?

What country wholly within the Temperate Zone? Within the Torrid Zone?

Name all the countries of North America, commencing at the most northern.

What three oceans wash the coasts of North America? Which is the coldest?

On which side are the greatest inlets from the ocean? Name them.

From what ocean does Hudson Bay receive water? Is the water warm, or cold? (See page 37, paragraph 18.)

Whence does the Gulf of Mexico receive its water? Is it warm, or cold?

How does each affect the temperature of the winds blowing over them?

What and where are the mountain systems of North America?

Between what chains is the Great Basin or Plateau of the United States?

Into what do the rivers flow which rise west of the Rocky Mountains?

Into what do those flow which rise on the east side?

Into what do the rivers flow which rise on the east side of the Alleghanies?

How far north does the basin of the Mississippi extend?

What two general slopes are between the Rocky and the Alleghany Mountains?

Mention the rivers of the Mississippi Basin,—of Hudson Bay Basin.

What two large rivers in the north-west? In the west?

What waters around Newfoundland? Greenland? Cuba? Lower California?

What land around the Gulf of St. Lawrence? Gulf of Mexico? Caribbean Sea?

In what cape does Greenland terminate? Nova Scotia? Florida? Lower California?

In what direction do these capes project?

Name all the capes on the Atlantic Coast? On the Pacific Coast?

What islands are washed by the Atlantic? Pacific? Caribbean Sea?

In the northern part of North America, what islands? Straits? Sounds?

What provinces in the south-eastern part of British America?

What lakes between the United States and Canada? What is their outlet?

Into what does the St. Lawrence flow?

What lakes are connected with the Arctic Ocean? With Hudson Bay?

What two high mountains north-west of British America?

What two in the south-western part?

* Canada East is now the Province of Quebec; Canada West, Ontario.

ROUTES OF TRAVEL.

In what directions and on what waters would you sail from Quebec to New York? From Boston to Iceland? From Washington to Dr. Kane's Open Polar Sea? What capes and islands would you pass in sailing from New York to New Orleans? New York to the Isthmus of Panama?

Sailing from the Isthmus to San Francisco, what direction would you take? On what water would you sail? What capes would you pass?

Refer to the scale of miles, and state the distance, in a straight line, from New York to Cuba. From Florida to Cuba. New Orleans to New York. New Orleans to the Isthmus of Panama. Washington to San Francisco.

What countries and parts of countries lie between the parallels of 40° and 50° north latitude? What parts of Europe and Asia lie between those parallels? (See margin of map.) What parts of America, Europe, and Asia lie between the parallels of 50° and 60°? What parts of the Eastern Hemisphere are directly east of the United States? West? What European Country is directly east of Washington? Of Nova Scotia and Newfoundland? Of the southern part of Labrador? Of the northern part of Labrador? What parts of North America lie between the same parallels as Sahara? China? Japan? Siberia?

REVIEW.

MOUNTAINS.

Where are they? In what directions do the ranges extend?

Mt. St. Elias? 17*	Alleghany?	Fremont's Peak?
Mt. Brown? 16	Mt. Hooker? 15	Rocky?
Mt. Fairweather? 15	Mt. Whitney?	Hecla?
Sierra Madre?	Sierra Nevada?	Cascade?
		Coast Range?

RIVERS.

Where do they rise? What courses do they take? Into what waters do they flow?

Missouri? 29†	Mackenzie?	Columbia?	Platte?
Mississippi? 30	Colorado?	Brazos?	Lewis?
Rio Grande? 18	St. Lawrence?	Nelson?	Ohio?
Athabasca? 15	Churchill?	Severn?	Red?

GULFS, BAYS, SOUNDS, AND CHANNELS.

Where are they? Into what waters do they open?

G. of St. Lawrence?	Hudson B.?	Chan. of Yucatan?
G. of California?	Baffin B.?	Fox Chan.?
B. of Honduras?	James B.?	Lancaster Sd.?
Mosquito Gulf?	Ungava B.?	Norton Sd.?
B. of Campeachy?	Frobisher's B.?	G. of Mexico?

STRAITS.

What lands are separated, and what waters are connected by them?

Hudson?	Windward?	Bellisle?	Davis?
Behring?	Wellington?	Florida?	Mona?
	Melville?	Barrow?	Banks?

LAKES.

Where are they? What are their outlets?

Superior?	Ontario?	Great Slave?	Huron?
Great Salt?	Winnipeg?	Great Bear?	Erie?
Athabasca?	Michigan?	Little Slave?	Itasca?

ISLANDS.

Where are they? By what waters are they surrounded?

Newfoundland?	Greenland?	West Indies?	Cuba?
Vancouver's?	Parry?	Bahamas?	Iceland?
Southampton?	Jamaica?	Bermudas?	Disco?
Queen Charlotte's?	Bank's Land.	Melville?	Sitka?
Cape Breton?	Porto Rico?	Anticosti?	Hayti?

CAPES.

Where are they? Into what waters do they project?

Hatteras?	Farewell?	Flattery?	May?
Mendocino?	St. Lucas?	Sable?	Cod?
	Blanco?	Race?	Icy?

* Numbers in lists of Mountains show height in thousands of feet. Mt. St. Elias, 17900 feet high.

† Those in lists of Rivers show length in hundreds of miles. Missouri, 2900 miles long.

DESCRIPTIVE GEOGRAPHY.

1. ***NORTH AMERICA*** is somewhat triangular in shape. Its widest part is from Newfoundland to Behring Strait; its narrowest, where Central America joins the Isthmus of Panama.

2. ***The Meridian of 97°*** west from Greenwich (or 20° from Washington) passes through the middle of North America, and near the western coasts of Hudson Bay and the Gulf of Mexico.

3. ***East of that Meridian*** are the great gulfs, bays, lakes, and lowlands; while west of it, are the great plateaus and mountain chains.

4. ***The Plateau or High Region*** extends over the western parts of British America and the United States and nearly the whole of Mexico and Central America. It is widest in the United States, under the 40th parallel of latitude.

5. ***From that Meridian*** the surface of the United States rises gradually to the Rocky Mountains, which are from 10,000 to 18,000 feet high.

6. ***Between the Rocky Mountains and the Sierra Nevada*** the surface is from 4,000 to 6,000 feet above the level of the sea, and is mostly dry and barren.

7. ***The Appalachian Chain*** near the Atlantic, and the Coast Range near the Pacific, are about one-fourth the height of the Rocky Mountains or the Sierra Nevada.

8. ***The Highest Peaks in North America are:*** over 18,000 feet high, Mt. Popocatepetl, in Mexico, and Mt. St. Elias, in Alaska; over 15,000 feet, Mounts Brown, Whitney, and Fairweather; over 14,000 feet, Pike's Peak, Mt. Shasta, and Mt. Tyndall.

9. ***The Gulfs, Bays, and Inlets*** from the Atlantic and Arctic Oceans help to form numerous peninsulas and islands. (*Which are partly formed by the Gulf of Mexico? By Hudson Bay? By Baffin Bay? By the Gulf of St. Lawrence?*)

10. ***The Great Watershed*** of North America is the Rocky Mountain chain, on the eastern side of which are the sources of nearly all the large rivers flowing into the Atlantic. (*Name them.*) On its western side are the sources of those flowing into the Pacific. (*Name them.*) The Mackenzie River receives its waters from both sides of the Rocky Mountains.

11. ***North America Lies*** in three zones and possesses every variety of climate, from the extreme cold of the Frigid to the excessive heat of the Torrid. The portion best adapted to the happiness and progress of the inhabitants is the intermediate or temperate zone, which includes the United States.

12. ***The Coldest Parts*** of North America are Greenland and the north-eastern part of British America. (*For the climate on the sides of high mountains in hot countries, see p. 38, paragraph 52.*)

13. ***The Climate of the Pacific Coast*** of the United States, British America, and Alaska is much milder than that of the Atlantic coast, in the same latitudes; because one is washed by the warm waters of the Japan current, the other by the cold currents from the Arctic Ocean (see p. 24, paragraphs 20 and 23).

14. ***Eastward from the Mackenzie River Valley*** to Baffin Bay is a vast, frozen, treeless region, while westward is a region of forests extending to the coast, where the climate is no more excessive than that of Maine or New Brunswick, hundreds of miles farther south on the Atlantic coast.

15. ***BRITISH AMERICA*** is separated from the United States by the 49th parallel of latitude, the Great Lakes, the St. Lawrence River, and the north-eastern portion of the Appalachian Chain.

16. ***Its General Slope*** is toward the north, the watershed on its southerly side extending generally along the northern boundary of Canada, and the north-western boundary of the United States.

17. ***All that portion which extends from the Mackenzie River Valley*** eastward to Baffin Bay and the Atlantic is a cold, barren region, covered nearly all the year with ice and snow, and is useful only as a hunting-ground. For a few weeks in summer, the snow and the top of the ground thaw, when mosses and other low plants appear; upon these the reindeer and some other animals feed.

18. ***The Fur-Bearing Animals*** are the beaver, marten, mink, bear, fox, wolf, and muskrat.

19. ***The Southern and Western Portions*** of British America contain extensive forests and vast tracts of prairie land well adapted to grazing and agriculture. On the pastures are herds of buffaloes, elks, and deer.

20. ***Hudson Bay Territory*** is north of the United States and Canada, and reaches from Alaska to Baffin Bay and Labrador; Rupert's Land, or New Britain, is all that portion which lies east of the Rocky Mountains. It is now all under the control of the Dominion of Canada.

21. ***BRITISH COLUMBIA***, lying north of Washington Territory, is a mountainous region, cold in winter and moist in summer. Its forests of pine, fir, and spruce are extensive.

22. ***Its Importance*** arose from the discovery of gold along Frazer River.

23. ***Its Chief Exports*** are lumber, gold, coal, furs and fish.

24. ***VANCOUVER ISLAND***, which is comprised in the Province of British Columbia, contains fertile valleys, fine timber, and rich mines of coal, besides copper and other ores.

25. ***The Climate*** is much milder than in the same latitudes on the Atlantic coast. Vancouver Island and British Columbia are under one governor (see page 37, paragraph 33).

26. ***The Capital*** and chief city of the Province of British Columbia is Victoria, in Vancouver Island.

27. ***THE PROVINCE OF MANITOBA*** is south of Lake Winnipeg, which receives the waters of the Red River of the North.

28. ***The Settlers*** are French, English, Scotch, and Americans. The French have Indian blood in their veins, and live mostly by hunting; the others, on the produce of their farms and pastures.

29. ***LABRADOR*** is cold, barren, and desolate along the coast, but well wooded in the interior.

30. ***The Inhabitants*** are almost exclusively Esquimaux; but, on the eastern shore, are settlements of seal-catchers, fur-traders, and Moravian missionaries. The eastern half of the peninsula is under the control of Newfoundland; the western, with Hudson Bay Territory and the Province of Manitoba, has been transferred to the government of Canada.

31. ***ALASKA***, now a Territory of the United States, is more than twice the size of France, but not more than one-tenth is habitable, on account of the extreme coldness of its climate.

32. ***Along the Southern Coasts*** the climate is tempered by the winds which blow over the warm waters of the Pacific. These winds are laden with moisture that supplies the rains and fogs for which that part of the Territory is celebrated.

Arctic Regions.—A Ship Caught in the Ice.

33. *Excepting along its Western and Northern Coasts, the Territory is Covered*, in some places heavily, with forests valuable for timber. The trees are chiefly evergreens—spruce, fir, edar, and hemlock. Its importance lies in its furs, fisheries, and orests.

34. *Seals, Whales, and Walruses* are plentiful in the waters est of Alaska. The fur seals on the islands of St. Paul and St. eorge are very numerous.

35. *Its Principal River* is the Yukon, which is said to rank ext to the Mississippi in size.

36. *The Inhabitants* are chiefly Esquimaux and Indians, who re engaged in fishing and hunting.

37. *Its Capital* is Sitka, situated on one of the isl ;nds which ine its coast.

38. *GREENLAND*, a vast island, or group of islands united y fields of ice, is rugged, mountainous, barren, and almost wholly overed with ice and snow.

39. *On the North-west Coast* laciers extend down into the sea, nd from them fragments break ff and float out as icebergs.

40. *Vegetation, in some arts*, appears in summer; ushes of birch, willow, and ountain ash are found, and a ew vegetables are raised in the outh.

41. *The Inhabitants*, about ,000 in number, are mostly Esuimaux, to whom the seal furishes food, fuel, and clothing.

42. *The Settlements of the anes*, who compose about oneghth of the population, are on e west coast.

43. *The Principal Settlements* e Julianshaab and New Hernt, which are south of the Arctic rcle, and Christianshaab and Uper'navik, north of it; the latter being the most northerly settlement of civilized man.

An Esquimaux.

44. *The Distinguished American Explorers* of the Arctic regions are Kane, Hall, and Hayes. The most northern point reached is a little beyond the latitude of 82 degrees.

45. *Greenland and Iceland* belong to Denmark, and are called Danish America.

46. *The Esquimaux*, except those of Alaska, are short—not over five feet in height—owing, probably, to the rigor of the climate and the nature of their food.

47. *The Color of their Faces* is brown; the result, chiefly, of their uncleanliness and their smoky huts (see page 46). Their bodies are of a much lighter color than their faces.

48. *The Huts of the Esquimaux* are partly under ground, and are constructed either of stone or of ice and snow.

49. *ICELAND* is larger than the State of Virginia.

50. *Its Surface* is very rough, containing volcanoes, fissures, lava fields, and innumerable boiling springs. Among the latter is the Great Geyser, which is among the wonders of the world (see page 27).

51. *Its Highest Mountains* are about 6,000 feet high; Mt. Hecla, its celebrated volcano, is about 5,000 feet above the sea level.

52. *It is Coldest* on the northern and north-eastern coasts, where snow falls even in summer; but on the southern and western coasts the air is tempered by the Gulf Stream. Here grasses grow abundantly and furnish food for large numbers of horses, cattle, and sheep.

53. *The Icelanders* are of Norwegian descent, and number about 64,000. For food, occupation, and exports, they depend mainly upon their fisheries and live stock.

54. *Herds of Reindeer*, imported originally from Norway, run wild over the uninhabited parts of the island.

55. *The Capital* and principal settlement is Reikiavik (rike-a-vik), which is on its south-western coast.

56. *Draw an outline of North America; then mark, in the following order—the Mountains—Rivers and Lakes—Bays and Capes—Countries.*

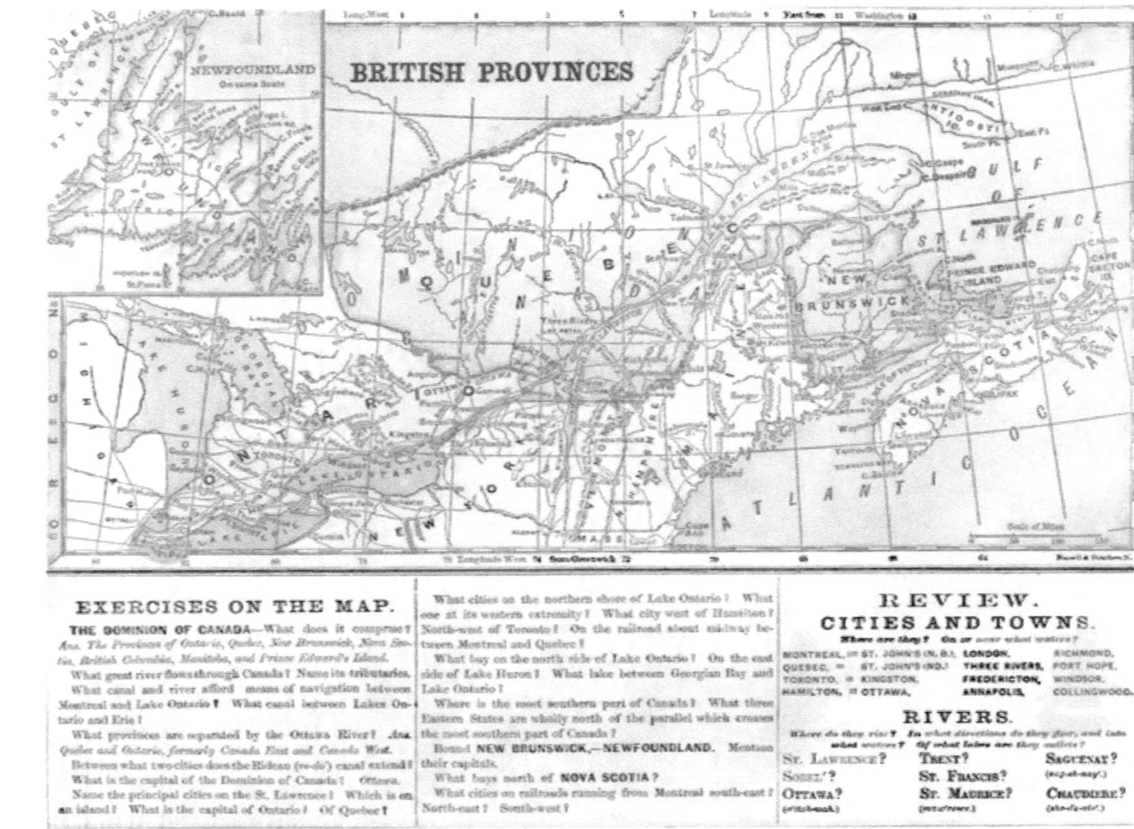

EXERCISES ON THE MAP.

THE DOMINION OF CANADA.—What does it comprise? *Ans. The Provinces of Ontario, Quebec, New Brunswick, Nova Scotia, British Columbia, Manitoba, and Prince Edward's Island.*

What great river flows through Canada? Name its tributaries.

What canal and river afford means of navigation between Montreal and Lake Ontario? What canal between Lakes Ontario and Erie?

What provinces are separated by the Ottawa River? *Ans. Quebec and Ontario, formerly Canada East and Canada West.*

Between what two cities does the Rideau (re-do') canal extend?

What is the capital of the Dominion of Canada? *Ottawa.*

Name the principal cities on the St. Lawrence? Which is on an island? What is the capital of Ontario? Of Quebec?

What cities on the northern shore of Lake Ontario? What one at its western extremity? What city west of Hamilton? North-west of Toronto? On the railroad about midway between Montreal and Quebec?

What bay on the north side of Lake Ontario? On the east side of Lake Huron? What lake between Georgian Bay and Lake Ontario?

Where is the most southern part of Canada? What three Eastern States are wholly north of the parallel which crosses the most southern part of Canada?

Bound **NEW BRUNSWICK**,—**NEWFOUNDLAND**. Mention their capitals.

What bays north of **NOVA SCOTIA**?

What cities on railroads running from Montreal south-east? North-east? South-west?

REVIEW.

CITIES AND TOWNS.

Where are they? On or near what waters?

MONTREAL,	ST. JOHN'S (N. B.),	LONDON,	RICHMOND,
QUEBEC,	ST. JOHN'S (ND.)	THREE RIVERS,	PORT HOPE,
TORONTO,	KINGSTON,	FREDERICTON,	WINDSOR,
HAMILTON,	OTTAWA,	ANNAPOLIS,	COLLINGWOOD.

RIVERS.

Where do they rise? In what directions do they flow, and into what waters? Of what lakes are they outlets?

ST. LAWRENCE?	TRENT?	SAGUENAY? (sag-eh-nay'.)
SOREL'?	ST. FRANCIS?	
OTTAWA? (ot'tah-wah.)	ST. MAURICE? (mo-reece'.)	CHAUDIERE? (sho-de-air'.)

DESCRIPTIVE GEOGRAPHY.

1. ***The British Provinces of Quebec*** **or** Lower Canada, and ***Ontario*** **or** Upper Canada, form **part** of the St. Lawrence basin, **which is** drained by the Great Lakes and the St. Lawrence **River.**

2. ***The Northern Watershed*** is along the northern boundary, and the principal slope is southeastward.

3. ***The Climate*** is excessive—colder in winter and hotter in summer than in the same latitudes in Europe. This is particularly so in the eastern province, where the winter lasts six or seven months of the year.

4. ***The Soil*** is generally good; grain, vegetables, fruits, and melons are raised.

5. ***Forests*** cover the greater part of the surface, and contain trees of every variety; lumbering and shipbuilding are extensively carried on.

6. ***The Principal Exports*** are wheat and timber, pot and pearl ashes.

7. ***Iron*** is abundant, and the region around Lake Superior and Lake Huron is famous for copper.

8. ***The Wild Animals*** include the elk, wolf, bear, wild cat, etc.

9. ***The First Inhabitants*** of Canada were Indians. Its discovery by Europeans was in 1535, by the French, under Jacques Cartier, who found Hochelaga, an Indian town, among rich cornfields (where Montreal now stands). The Indians were friendly, until the French carried off one of their kings. For many years the settlers suffered much from the climate and the natives.

10. ***These Provinces*** remained in possession of the French more than two centuries, or until their cession to Great Britain in 1763, which followed the capture of Quebec by General Wolfe.

11. ***The Inhabitants*** of Quebec are mostly of French origin; of Ontario, British. The French language is spoken in the former, and English in the latter.

12. ***The Dominion of Canada Comprises*** all the British Provinces of North America, except Newfoundland. They have a parliament and a governor-general.

13. ***Each of these Provinces*** has its lieutenant-governor and legislative body, but all are subject to Great Britain.

14. ***The Capital*** is Ottawa, on the Ottawa River; the metropolis is Montreal, on Montreal Island; and the most strongly fortified city in America is Quebec; both are situated on the St. Lawrence River, and are in more southern latitudes than the European cities of London, Paris, or Vienna.

15. ***The other Important Cities*** are Toronto, Hamilton, and Kingston, which are situated on Lake Ontario. Toronto is in nearly the same latitude as Portland, Idaho, and Eugene City (United States), Nice and Florence (Europe).

16. ***the Natural Objects of Interest*** are the Falls of Niagara and of Montmorency, the Thousand Isles, the Rapids in the St. **Lawrence, and the** high, perpendicular cliffs of the Saguenay **(*Sag-eh-nay'*) River.**

17. ***NEW BRUNSWICK:*** Its highlands are in the north and north-west; its principal river is the St. John's, on which, at the head of steam navigation, is Fredericton, the capital of the province.

18. ***Its dense Forests,*** which cover the greater part of **its surface,** and its numerous inlets from the Bay of Fundy **and the Gulf of** St. Lawrence, furnish the principal exports—lumber, **fish, and fish-oil.**

19. ***The Metropolis*** is St. John, noted for its fine harbor, shipbuilding, lumber trade and fisheries. The other important towns are St. Andrew's and Liverpool.

20. ***The Bay of Fundy is remarkable*** for its extraordinary tides, which rise suddenly,—in some places, to the height of over 60 feet.

21. ***New Brunswick and Nova Scotia,*** originally a French colony, under the name of Acadia, or New France, was ceded to Great Britain in 1713.

22. ***NOVA SCOTIA*** is a peninsula, which, with Cape Breton Island, east of it, forms one of the British Provinces.

23. ***Its Forests*** are extensive, rivers short, and climate excessive, with heavy fogs along the Atlantic coast (see p. 25, par. 23—26).

24. ***The Valleys*** in Nova Scotia and New Brunswick are fertile, and yield good crops of grain, potatoes, turnips, &c.

25. ***Its Exports*** are the same as those of New Brunswick, with coal and grindstones, fish and fish oil.

26. ***Considerable Attention is paid*** to commerce and shipbuilding.

27. ***Halifax,*** the capital and metropolis, has one of the finest harbors in the world.

28. ***Among the other Important Towns*** are Yarmouth, Windsor, Pictou', and Sydney.

29. ***PRINCE EDWARD'S ISLAND,*** situated in the southern part of the Gulf of St. Lawrence, is noted for its indented coast, fine farms and pastures, and its healthful climate.

30. ***Its Capital*** is Charlottetown.

31. ***NEWFOUNDLAND,*** an island larger than either Scotland or Ireland, is a dreary region, composed of barrens, marshes, and lakes. It is in the same latitude as France (see page 36, paragraph 10).

32. ***It is Noted*** for long winters and dense sea fogs.

33. ***Its Waters*** swarm with fish; chiefly cod, salmon, herrings, and mackerel. Seals are numerous. Its cod-fisheries on the banks, south and south-east of the island, are world renowned. Nearly all the inhabitants are engaged either in catching or curing fish.

34. ***Its Exports*** are dried fish, fish-oil, seal-skins, and seal-oil.

35. ***Its only Town*** is St. John, the capital, which, with most of the villages, is situated on the south-east coast.

36. ***The Population*** of New Brunswick, Nova Scotia, Newfoundland, and Prince Edward's Island consists chiefly of emigrants from the British Isles, or their descendants.

37. ***Newfoundland Forms*** a distinct colony, under a governor appointed by the British Crown, with a local legislature.

38. ***It was Discovered*** in 1497 by John Cabot and his son Sebastian, who sailed from England. The natives were wild and unfriendly, clothed with the skins of animals, and painted with reddish clay.

Newfoundland is said to have been first discovered in the 11th century, by the early colonists of Iceland and Greenland. Iceland was settled in the 9th century, by Norwegians.

BRITISH PROVINCES.

NAMES.	POPULATION.	NAMES.	POPULATION.
Ontario	2,135,308	Newfoundland........	150,000
Quebec....	1,422,546	Prince Edward's Island	95,000
Nova Scotia	387,800	British Columbia......	50,000
New Brunswick.....	306,449	Manitoba..........	12,297

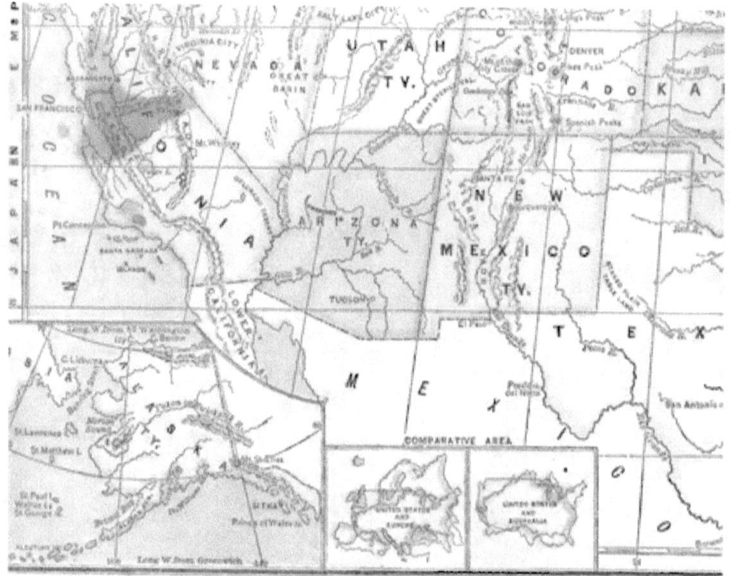

EXERCISES ON THE MAP.

Bound the **UNITED STATES.** What is the capital of the Republic ?

What is the general direction of the Atlantic Coast ?

What is the direction of the coast of Maine ? Massachusetts ? Of that from Long Island to Cape Charles ? Cape Charles to Cape Hatteras ? Cape Hatteras to the northern boundary of Florida ? Of the eastern coast of Florida ?

What is the direction of the Pacific coast from Cape Flattery to Cape Mendocino ? Cape Mendocino to Lower California ?

What part of the northern boundary is furthest north ? On what parallel of latitude is that part ? What European city near that parallel ? (See Margin of Map.)

Name the lakes and rivers on the northern boundary ?

Where is the most southern part of the northern

From what States and Territories can you ent
crossing a lake or a river ?

What State is in the central part of the Union ?

Which is the most north-eastern State ? What S
further north than Maine ? That the most southe

What States are entirely separated from British A

What State partly separated from it by lakes an

Mention the lakes and rivers between New York

What States and parts of States form peninsulas

What fourteen States are washed by the waters

What five by the Gulf of Mexico ? What States ar

What States on the east bank of the Mississippi? On the west bank?
What States on the north bank of the Ohio? On the south Bank?
What States border on the Missouri River? What States and Territories bor-
r on Mexico?
What States are greatest in extent from north to south? From east to west?
What is the greatest latitude of the northern boundary? Of the southern
undary?
Through how many degrees of latitude does this country extend?
What is the longitude of its most eastern boundary, reckoning from Washing-
n? Of its most western? Of its most eastern boundary, reckoning from
eenwich? Of its most western?
Through how many degrees of longitude does this country extend?

What cities in Europe are in more northern latitudes than Quebec? Than Montreal? Than Massachusetts?

What parallel of latitude passes over Philadelphia? What Capitals in the United States on or near that parallel? What Capitals in the Eastern Hemisphere on or near it?

What cities, of both hemispheres, are situated in the same or nearly the same latitude as New York? San Francisco? Nashville? Savannah? New Orleans?

Name the places in the United States which have the same latitude as Paris? Florence? Madrid? Pekin? Lisbon? Gibraltar? Morocco?

Name the cities, of both hemispheres, which are situated between the parallels of 28° and 32° north latitude? Between 32° and 36°? Between 36° and 40°? Between 40° and 44°? Between 44° and 48°?

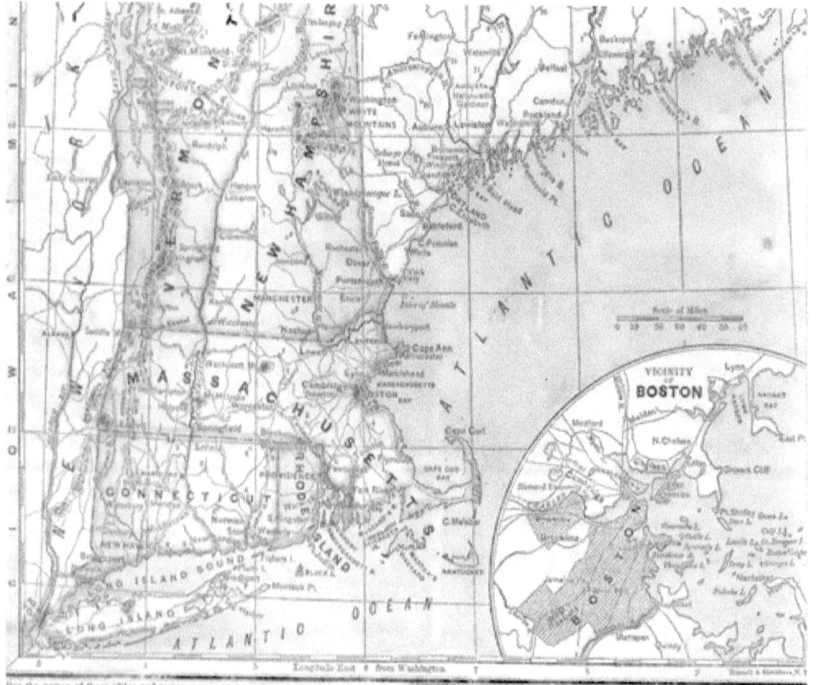

For the names of these cities and towns represented on the maps by numbers, see Appendix.

Monteith's P.

EXERCISES ON THE MAP.

Name the Eastern States, commencing with the largest.

Which have sea coast? Which border on the British Provinces?

Mention the bays on the coast, commencing with the eastern part.

What capes on the coast of Maine? Of Massachusetts?

What four boundary rivers have the Eastern States? What boundary lake?

Which is the most mountainous of the eastern States? Name the mountains.

Bound **MAINE.** What is its capital? What single mountains in Maine?

Into what do the rivers in the northern part of Maine flow?

Into what do most of the rivers in the State flow?

What two general slopes has Maine?

What lakes in the region of the watershed?

In what part of Maine are its largest cities and towns?*

Which are situated on the sea coast? On rivers?

Bound **NEW HAMPSHIRE.** What is its capital? What mountains in the State?

Which is the highest peak of the White Mountains?

What is the largest river in New Hampshire? Largest lake?

What lake in the north? On the north-eastern boundary?

In what part of the State are its largest cities and towns?

What three on the Merrimac? In the south-eastern part of the State?

What two towns in the south-west? What town on the Connecticut west of the White Mountains?

Bound **VERMONT.** What is its capital?

What mountains constitute the watershed of Vermont?

In what directions does the land slope?

Where is the land more elevated, at the center or sides?

Into what do the rivers of Vermont flow?

Name the principal cities and towns in the State.

Bound **MASSACHUSETTS.** What is its capital?

What mountains extend through the State? What large river flows through the western part? The north-eastern part?

What river from Massachusetts flows into Rhode Island? What two into Connecticut?

What two islands south-east of the State?

What cities in Massachusetts on the Merrimac? What cities in the east?

What cities in the south-east? In the west? What on the Connecticut? On the Blackstone River?

What single mountains in Massachusetts?

Bound **CONNECTICUT.** What is its capital?

What rivers flow through the State? Into what do they flow?

In what state is the source of the Connecticut River? Of the Housatonic?

In what direction does the surface of Connecticut slope?

What city in the north? East? What two in the southern part?

Bound **RHODE ISLAND.** What are its capitals?

What large bay in the State? What river flows into it?

In what State is the source of the Blackstone River?

Mention the principal cities in Rhode Island.

On what island is Newport situated? Ans. Rhode I.

Draw a map of the Eastern States. (*See Appendix.*)

ROUTES OF TRAVEL.

On what waters and near what islands would you sail from Portland to New Haven? In what directions, and on what waters, from Hartford to New York? New York to Fall River?

At what cities do several railroads meet?

What directions would you take and what cities would you pass in traveling by railroad from Boston to Hartford? Boston to Albany? Boston to Portland? Portland to Montreal? Montreal to Rutland? Rutland to Boston?

Refer to **the Scale of Miles and state** the distance in a straight **line** from **Boston to New York,—Boston to** Albany,—Boston to Portland,—Portland **to Mt. Washington,—Mt. Washington** to Montreal. (For exercises **on the margins of the map, see p. 102.**)

If the State in which you reside be represented on this map, the following will be additional exercises:

Give the direction from you to Boston,—Albany,—New York,—Newport,—New Haven,—Burlington,—Augusta,—Mt. Washington. Point toward each.

Mention all the cities and towns in the northern part of your State,—in the eastern,—southern,—western,—central part.

How many miles from you to the capital of your State? To its largest city? (See population of Cities in *Review*.)

What is the population of each of the largest cities in your State?

Name all the cities and towns on the map within fifty miles of your residence. Name those in both hemispheres that are in the same latitude as the city in or near which you reside. (*See Map of U. S.*)

Draw a map of your State.

REVIEW.

CITIES AND TOWNS.

Where are they? On or near what waters?

BOSTON, 250	SALEM, 24	CONCORD, 12	MONTPELIER,
PROVIDENCE, [illegible]	MANCHESTER, [illegible]	NEWPORT, [illegible]	PITTSFIELD,
NEW HAVEN, [illegible]	LAWRENCE, [illegible]	NEW LONDON,	AUGUSTA,
LOWELL, 41	BANGOR, [illegible]	NASHUA, [illegible]	BELFAST,
HARTFORD, [illegible]	SPRINGFIELD, [illegible]	PORTSMOUTH, [illegible]	LUBEC,
PORTLAND, [illegible]	NORWICH, [illegible]	DOVER, [illegible]	MIDDLEBURY,
CAMBRIDGE, [illegible]	FALL RIVER, [illegible]	RUTLAND,	GARDINER,
LYNN, [illegible]	NEWBURYPORT, [illegible]	BATH, [illegible]	ST. ALBANS,
TAUNTON, [illegible]	BRIDGEPORT, [illegible]	BURLINGTON,	EASTPORT,
NEW BEDFORD, [illegible]	GLOUCESTER, [illegible]	WATERBURY,	BENNINGTON,
WORCESTER, 41 (woos'ter)	CALAIS, (kal'is)	HAVERHILL, (hay'ver-il)	MIDDLETOWN,
			WINCHESTER.

MOUNTAINS.

Where are they? In what directions do the ranges extend?

MT. WASHINGTON? 6¼	MT. SADDLEBACK? 4	MT. KATAHDIN? 5¼
WHITE MTS.? 5¼	MT. EVERETT? 2¾	SADDLE MT.?
GREEN MTS.? 4	MT. WACHU'SETT? 2	CAMEL'S HUMP? 4

RIVERS.

Where do they rise? Between what and through what States do they flow? Into what waters do they flow?

PENOBSCOT? 4	AROOSTOOK?	ST. FRANCIS?	SOREL?
KENNEBEC? 3	WOOLASTOOK?	OTTER CREEK?	ONION?
ANDROSCOGGIN? 2	LA MOILLE?	SALMON FALLS?	SACO? (saw'ko)
MERRIMAC?	ST. JOHN?	BLACKSTONE?	
CONNECTICUT? 5 (kon-net'i-kut)	HOUSATONIC? (hoo-sa-ton'ik)	ST. CROIX? (sent kroy')	THAMES? (temz)

LAKES.

Where are they situated? What are their outlets?

MOOSEHEAD?	MEMPHREMA'GOG?	UMBA'GOG?
WINNIPISEOGEE? (win-ne-pe-saw'ke)	CHAMPLAIN?	CHESUN'COOK?
	CONNECTICUT?	SCHOODIC?

BAYS.

Where are they? Into what waters do they open?

NARRAGANSETT?	FRENCHMAN'S?	BUZZARD'S?	CAPE COD?
PENOBSCOT?	LONG ISLAND SD.?	CASCO?	FUNDY?

* Railroads are shown by dotted lines. The pupils may include in their answers to all such questions the names only of those Cities and Towns which appear on the maps in black letters.

The numbers in Cities and Towns, show the population, in thousands, according to the census of 1870: Boston, 250,526.

Numbers in lists of Mountains show heights, in thousands of feet: White Mts., 5000 feet; those in list of Rivers, lengths in hundreds of miles: Connecticut R., 500 miles.

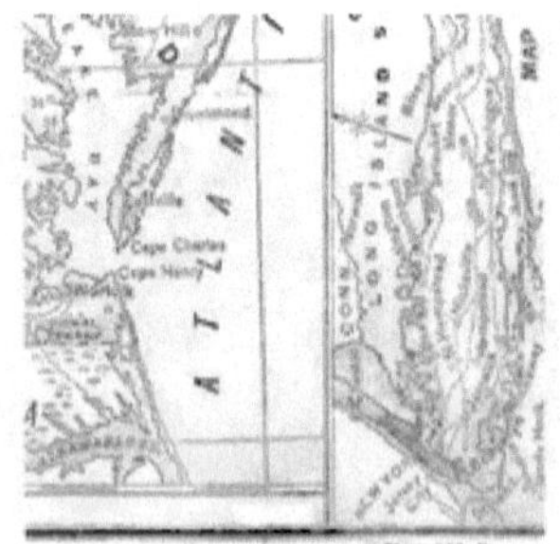
Cape Charles
Cape Henry

EXERCISES ON THE MAP.

Bound **NEW YORK.** What part borders on the Atlantic ocean?
What three mountain ranges in New York?
In what direction does the land west of the Adirondacks slope?
Name the rivers on that slope. Into what do they empty?
What large river has its source on the east side of the Adirondacks?
What river flows through the western part of the State? The eastern?
What large tributary has the Hudson?
What tributary of the Susquehanna is in New York?
What lakes in the center of the State? By what are they drained?
Is the level of Lake Ontario higher or lower than that of these lakes?
What two islands in the south-eastern part of the State?
What water north of Long Island? South?
What large city in Long Island?
Name the cities and towns on the Hudson, *commencing* at its mouth.
What towns in the northern part of the State?
What cities on the Central Railroad, between Albany and the Niagara River? What city at the eastern extremity of Lake Erie?
What river and canal cross each other at Rochester?
What Canadian town opposite Ogdensburg? What city at the mouth of the Oswego River? (Canada East is now called Quebec; Canada West, Ontario.)
Bound **PENNSYLVANIA.** Name its mountains and large rivers.
In what direction do its mountain chains extend?
What two large tributaries of the Susquehanna in Pennsylvania?
On which side are its large tributaries?
By what rivers is the land east of the Susquehanna drained?
What cities and towns between the Susquehanna and Delaware Rivers.
What city on the Delaware opposite Philadelphia?
What city in the south-west? What three rivers form a junction at Pittsburg? What city in Pennsylvania on Lake Erie?
Has Pennsylvania any sea coast? What rivers and bays form outlets to the ocean? Where are the great coal regions? Where is the oil region?
Bound **NEW JERSEY.** What part is mountainous?
What large cities in the north-east? In the west? Island east?
What cape in the south? What on the eastern coast?
Bound **DELAWARE.** What city in its northern part?
With what city in Pennsylvania is it intimately connected?
What two lines of communication between them?
What city near the center of the state?
What cape on its coast? What cape opposite Cape Henlopen?
Bound **MARYLAND.** What bay almost divides the State into two separate parts? On which side of the bay is the greater portion?
Where are the mountainous districts of Maryland?
What two cities on the western shore of Chesapeake Bay?
What tract of land on the Potomac belongs exclusively to the General Government?
Bound the **DISTRICT OF COLUMBIA.** What important city does it contain?
Bound **VIRGINIA.** What mountains extend through the State?
What mountains on the south-western boundary?
What rivers east of the Alleghany Mountains?
Bound **WEST VIRGINIA.** What is its capital?
By what rivers is the land west of the Alleghany Mountains drained?
Name the principal cities in Virginia. In West Virginia.
Where is the oil region? Where are the salt works? The warm and the Sulphur Springs?

ROUTES OF TRAVEL.

What directions would you take, and what cities would you pass, in travelling by railroad from New York to Rome? From Rome to Niagara Falls? From Rochester to Cleveland? From New York to Dunkirk?

What cities do you pass, and what rivers do you cross, in traveling by railroad from New York to Washington?

From Philadelphia to Pittsburg? From Pittsburg to Niagara Falls?
From Wheeling to Baltimore? From Dover to Trenton?

What cities and capes would you pass in sailing from Albany to Philadelphia? From Philadelphia to Baltimore?

What two railroad routes from Jersey City to Elmira? Richmond to Lynchburg? Ogdensburg to New York?

What is the shortest route from New York to Richmond? From New York to Montreal? From Harrisburg to Norfolk?

What canal affords means of navigation between Lakes Erie and Ontario? *Ans.* Welland Canal.

Refer to the Scale of Miles, and give the *distance*, in a straight line, from Philadelphia to Harrisburg. From New York to Washington. Washington to Richmond. Baltimore to Wheeling. New York to Montreal. (For exercises on the margins of the map, see page 102.)

If the State in which you reside be represented on this map, the following will be additional exercises:

Give the direction from you of Washington,—Philadelphia,—New York,—Albany,—Montreal,—Niagara Falls,—Richmond. Point toward each.

Mention all the cities and towns in the northern part of your State. In the eastern,—southern,—western,—central part.

How many miles from you to the capital of your State?

Name all the places on the map within fifty miles of the city or town in or near which you reside. Name the cities of both hemispheres that are in the same latitude as that in or near which you reside. (*See Map of U. S.*)

Draw a map of your State.

What is the population of the largest cities in your State?

REVIEW.

CITIES AND TOWNS.

Where situated? On or near what waters?

NEW YORK, 942*	RICHMOND, 51	OSWEGO, 20	ELMIRA,
PHILADELPHIA, 674	JERSEY CITY, 82	KINGSTON,	WATERTOWN,
BROOKLYN, 396	ALLEGHENY, 53	NEWBURG,	OGDENSBURG,
BALTIMORE, 267	SYRACUSE, 43	NORFOLK,	LYNCHBURG,
BUFFALO, 118	READING, 34	POUGHKEEPSIE,	
NEWARK, 105	UTICA, 28	CAMDEN,	PLATTSBURG,
ALBANY, 69	WILMINGTON, 31	WHEELING,	ROME,
WASHINGTON, 109	PATERSON, 33	HARRISBURG,	DOVER,
PITTSBURG, 86	PETERSBURG, 19	POTTSVILLE,	PARKERSBURG,
ROCHESTER, 62	LANCASTER, 20	ERIE,	DUNKIRK,
TROY, 46	TRENTON, 23	SCRANTON,	CHARLESTON.

MOUNTAINS.

Where are they? In what directions do they extend?

ALLEGHANY? CUMBERLAND? LAUREL RIDGE? CATSKILL?
ADIRONDACK? BLUE RIDGE? CHESTNUT RIDGE? HIGHLANDS?

RIVERS.

Where do they rise? Between what, or through what States do they flow? Into what waters do they flow?

SUSQUEHANNA? 4†	SHENANDOAH?	PAMUNKY?	OSWEGO?
ST. LAWRENCE? 7½	DELAWARE? 3	BIG SANDY?	OHIO? 9
MONONGAHELA? 3	ALLEGHANY? 3	HUDSON? 3	YORK?
RAPPAHANNOCK 1½	GENESEE? 1½	POTOMAC? 4	MOHAWK? 1½
GREAT KANAWHA? 3	JUNIATA?	NIAGARA?	RACKET?

LAKES.

Where are they? What are their outlets?

CHAMPLAIN? 130‡	GEORGE? 36	CAYUGA? 38	ERIE? 250
ONTARIO? 190	ONEIDA? 20	OWASCO? 11	HURON? 260
SKENEATELES? 16			SIMCOE? 30

---

* The population in thousands, according to Census of 1870.
† Numbers in list of rivers show length in hundreds of miles. Ohio River—900 miles long.
‡ Those in list of lakes show whole length. Lake Champlain—130 miles long.

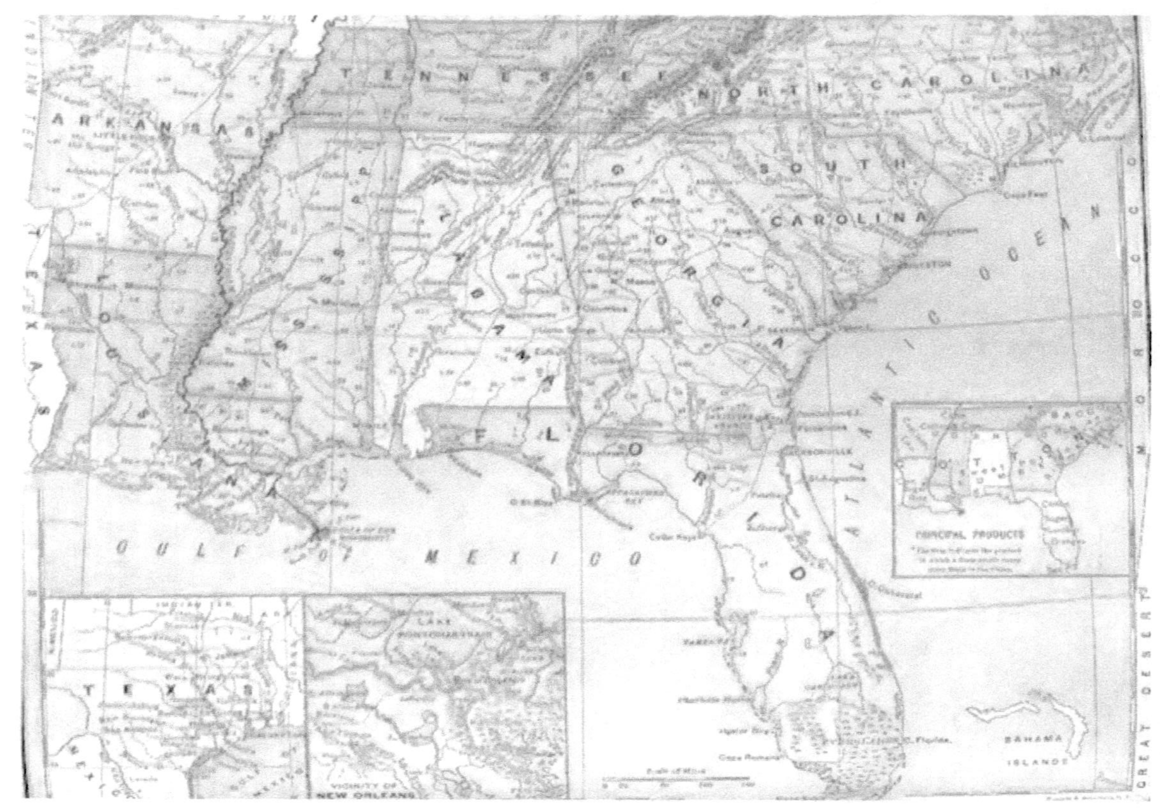

TENNESSEE
NORTH CAROLINA
ARKANSAS
SOUTH CAROLINA
GEORGIA
ALABAMA
MISSISSIPPI
LOUISIANA
FLORIDA
TEXAS
ATLANTIC OCEAN
GULF OF MEXICO
PRINCIPAL PRODUCTS
BAHAMA ISLANDS
LAKE PONTCHARTRAIN
VICINITY OF NEW ORLEANS
GREAT DESERT
MOROCCO

EXERCISES ON THE MAP.

Which of these States border on the Gulf of Mexico? On the Atlantic?

Which have no sea coast? What river affords them communication with the gulf?

What three States have their northern boundaries near the parallel of 36½ degrees north latitude? On the parallel of 35 degrees?

What southern State is a peninsula? By what waters is it embraced?

Where are the mountainous districts of these States?

Name all the bays, sounds, and inlets between Virginia and Cape Sable.

Name those from Cape Sable to Texas.

Name the capes on the Atlantic Coast. On the Gulf Coast.

Name all the boundary rivers of these States.

What is the character of the coast of the Southern States?

What rivers drain the Atlantic Slope? The Gulf Coast?

What rivers drain this part of the Mississippi Valley?

What States are drained by the Gulf and the Mississippi River?

What State is drained entirely by the Mississippi and its tributaries?

What part of Alabama is not drained by rivers flowing into the gulf?

What part of North Carolina is mountainous? Of Tennessee? Of Georgia? Of Arkansas?

In what mountains do many of the rivers of these States rise?

What two States are partly on the Atlantic, and partly on the Gulf Slope?

Bound **NORTH CAROLINA.** Mention its capital.

What city in the north-east? South?

What two cities in the east?

What capes on the coast? Are they on the main land?

What two sounds enclosed by the islands east of North Carolina?

Bound **SOUTH CAROLINA.** Mention its capital.

What are the principal cities in South Carolina?

What rivers in this State flow into the Atlantic ocean?

What two rivers form the Santee? The Congaree?

What rivers of South Carolina rise in North Carolina?

In what direction do all the rivers in South Carolina flow?

Bound **GEORGIA.** What is its capital?

What two cities on the west bank of the Savannah?

What city near the center of the State? In the west?

What three towns in the northern part of Georgia?

What rivers of Georgia flow into the Atlantic? Into the Gulf? What rivers form the Altamaha? The Appalachicola?

Bound **FLORIDA.** What is its capital?

What city in the north-east? What three in the west?

What rivers in Florida flow into the Gulf? Into the Atlantic?

What islands south of Florida? South-east?

What bay receives all the drainage of Alabama?

What river flows into Mobile Bay? What form the Mobile River?

What three large tributaries has the Alabama River?

What tributary has the Tombigby?

What large river flows through the northern part of Alabama?

Bound **MISSISSIPPI.** What is its capital?

What cities on the Mississippi River? What on the Tombigby?

What rivers in this State flow into the Mississippi?

What tributaries has the Yazoo River?

Bound **LOUISIANA.** What is its capital?

What is the largest city in the Southern States? (*See Review.*)

Where is the most southern part of the State?

Name the rivers of Louisiana.

Bound **TENNESSEE.** What is its capital?

What city in the south-west? North-east?

What two large rivers flow through Tennessee?

What mountains in the eastern part of the State?

Bound **ARKANSAS.** What is its capital?

Name the rivers in Arkansas. The mountains.

ROUTES OF TRAVEL.

In what directions, and in what water, would you sail from Virginia to Charleston? From Charleston to New Orleans? From Augusta to Newbern? From Nashville to Natchez?

What directions would you take, and what cities would you pass, in traveling by railroad from Jackson to Savannah? From Savannah to Montgomery? Pensacola to Rome? Selma to New Orleans? Memphis to Knoxville? Chattanooga to Savannah?

Refer to the Scale of Miles, and give the distance from Savannah to Charleston,—Savannah to Chattanooga,—Chattanooga to Memphis,—Memphis to New Orleans,—Vicksburg to Montgomery,—Montgomery to Mobile. (For exercises on the margins of the map, see page 102.)

If the State in which you reside be represented on this map, the following will be additional exercises:

Give the direction from you of Savannah,—New Orleans,—Memphis,—Charleston,—Newbern,—Pensacola,—Atlanta. At the same time point toward each.

Name all the cities and towns in the northern part of your State,—in the eastern,—southern,—western,—central part.

How many miles from you to the capital of your State?

Name all the places on the map within fifty miles of the city or town in or near which you reside. Draw a map of your State.

What is the population of the largest cities in your State?

What places are in the same latitude as the city in or near which you reside? (*See Map of U. S.*)

CITIES AND TOWNS.

Where situated? On or near what waters?

NEW ORLEANS,	WILMINGTON,	SELMA,
CHARLESTON,	ATLANTA,	MILLEDGEVILLE,
MOBILE,	COLUMBIA,	FLORENCE,
MEMPHIS,	MACON,	SHREVEPORT,
SAVANNAH,	NEWBERN,	JACKSON,
NASHVILLE,	PENSACOLA,	ROME,
NATCHEZ,	VICKSBURG,	COLUMBUS, (MISS.)
AUGUSTA,	LITTLE ROCK,	CHATTANOOGA,
MONTGOMERY,	KEY WEST,	GEORGETOWN,
COLUMBUS (GA.),	TALLAHASSEE,	JACKSONVILLE,
RALEIGH,	KNOXVILLE,	ATHENS,
BATON ROUGE,	BEAUFORT,	ST. AUGUSTINE,

BAYS, SOUNDS, AND INLETS.

Where situated? Into what waters do they open?

APPALACHEE?	BLACK?	MOBILE?	ONSLOW?
ALBEMARLE?	RALEIGH?	TAMPA?	PAMLICO?

RIVERS.

Where do they rise? What courses do they take? Into what waters do they flow?

RED? 25	CAMBAHEE?	ROANOKE?
TENNESSEE? 8	APPALACHICOLA?	WATEREE?
WHITE? 6	TALLAPOOSA?	ST. FRANCIS?
CUMBERLAND? 6	BIG BLACK?	ST. MARY'S?
CHATTAHOOCHEE? 5	OCMULGEE?	FLINT?
GREAT PEDEE? 4	SUNFLOWER?	NEUSE?
SAVANNAH? 4	YADKIN?	BROAD?
SANTEE? 4	BLACK WARRIOR?	PEARL?
TOMBIGBY? 4	ALABAMA?	YAZOO?
CAPE FEAR? 3	CAHAWBA?	COOSA?
LITTLE PEDEE?	CONGAREE?	ST. JOHN'S?
MOBILE?	FRENCH BROAD?	ED'ISTO?
OGEECHEE?	ARKANSAS?	OCONEE?
ALTAMAHA?	SUWANEE?	SABINE? 8

CAPES.

Where situated? Into what waters do they project?

CANAVERAL?	LOOKOUT?	SABLE?	FEAR?
HATTERAS?	ST. BLAS?	ROMANS?	

EXERCISES ON THE MAP.

What State has the greatest extent of lake coast?

What States border on Lake Superior? Lake Michigan? Lake Erie?

What bay opens into Lake Michigan? What two into Lake Huron?

Which of these lakes is the most elevated? What river receives their surplus water? (See illustration, page 21.)

Through what strait does water flow from Lake Michigan? Through what river from Lake Huron? From Lake St. Clair?

By what three rivers are these States mostly drained?

What land is drained by the great lakes?

In what Western States are railroads most numerous?

What three States on this map are on the west side of the Mississippi? What three on the east side, or left bank? What three on the right bank of the Ohio?

Name the boundary rivers of these States.

Bound **OHIO.** What is its capital?

In what two general directions does the land of Ohio slope?

What rivers on its northern slope? Southern?

What two cities in Ohio, on Lake Erie?

If you cross Lake Erie from these cities, what land would you reach?

What city near the western extremity of the lake?

Mention the largest river shown on this map.

What cities in Kentucky opposite Cincinnati?

What cities south-west of the capital of Ohio?

What city on the Maumee? On the Muskingum?

Bound **INDIANA.** What is its capital? What rivers in Indiana?

What corner of the State is washed by Lake Michigan?

What large city in the south-west?

What large cities on the Wabash River? Ohio River?

What city in the north-eastern part of Indiana?

What two cities in Ohio are nearest Indianapolis?

What two in Kentucky nearest Indianapolis?

What large city in Indiana nearest Louisville?

Bound **ILLINOIS.** What is its capital? What rivers flow into the Mississippi? Into the Wabash?

What city in Illinois on Lake Michigan? What other large cities in the northern part of the State?

What cities on the Wabash River? What city south-east of Peoria?

What city at the mouth of Rock River?

What city in Iowa opposite Rock Island?

What city on the Mississippi west of Springfield? South-west?

What large city in Missouri near Alton?

In what parts of Illinois, Wisconsin, and Iowa are celebrated lead mines?

Bound **MICHIGAN.** What is its capital?

What mines near Lake Superior? Name the rivers of Michigan.

Into how many parts is Michigan divided? Which is the larger part?

What large city in the south-east? What city in Ohio nearest Detroit?

What city on the Kalamazoo River? What two on the Grand? On the Saginaw River?

Bound **WISCONSIN.** What is its capital? Name its principal cities.

What cities on Lake Michigan? On Lake Winnebago?

What are its rivers? What lake in the eastern part of the State?

Bound **MINNESOTA.** What is its capital? Name its principal cities.

On what river are they? What river has its source in Itasca Lake?

Name the lakes in Minnesota.

What rivers in Minnesota flow into the Mississippi? What boundary river north? North-west?

Bound **IOWA.** What is its capital? Name its principal cities and towns.

In what direction does the land slope? Name the rivers of Iowa. Into what do they flow? What city in the lead district?

Bound **MISSOURI.** What is its capital? Name its principal cities.

What rivers in Missouri? Mountains? Mines?

Bound **KENTUCKY.** What is its capital? Name its principal cities.

By what rivers is Kentucky drained? Into what do they flow?

In what mountains do many of its rivers rise?

For "REVIEW" and "ROUTES OF TRAVEL" see Appendix.

EXERCISES ON THE MAP.

What is the largest State in the Union?

Bound **TEXAS**. What is its capital? See page 64.

In what direction and into what waters do the rivers of Texas flow?

What rivers drain the northern part of Texas? The western part?

What rivers flow into the Gulf of Mexico?

What city on the Colorado? What south-west of Austin?

What town in the most southern part of Texas?

What Mexican town opposite Brownsville?

What river separates Brownsville from Matamoras?

What two states of Mexico are separated from Texas by the Rio Grande?

What Territory between Texas and Kansas?

Bound **INDIAN TERRITORY**.

Mention the rivers that drain Indian Territory. Which of them empty into the Mississippi?

What State is in the central part of the Union? *Ans.* Kansas.

What city is in the central part of the Union? *Ans.* Topeka.

By what is **KANSAS** bounded on the north? South? West?

What rivers drain the northern part of Kansas? The southern part?

In what direction do the rivers in Kansas flow?

By what is **NEBRASKA** bounded on the north? South? West?

Mention the rivers of Nebraska. In what direction do they flow?

By what is **DAKOTA** bounded on the north? South? West?

What is the capital of Dakota?

What great river flows through Dakota?

What rivers flow into the western side of the Missouri? Into the eastern?

What lake in the northern part? Into what does its outlet empty?

Into what does the Red River of the North flow? (See Map of North America.)

What river rises a few miles west of Lake Mini Wakan?

Into what does the Dakota River discharge its waters?

What State in the Union is the second in size?

Bound **CALIFORNIA**. What is its capital?

In what direction does the State extend?

On which side is its sea coast?

What joins California on the north? On the east? South? West?

What Mountains in the eastern part of the State? Western?

What two rivers drain the valley between the Sierra Nevada and the Coast Mountains?

In what direction does the Sacramento flow? The San Joaquin? (sahn wah-keen'.)

What bay receives their waters?

What large river on the south-eastern boundary?

What river flows through the north-western part of California?

Of what lake is it the outlet?

What single mountain in the northern part of the State near Sacramento River?

What single mountain in the north-west near the coast?

What lakes in California have inlets but no outlets?

What important city at the entrance to San Francisco Bay?

What city south-east of San Francisco?

What town in the north-western part of the State?

What town in the southern part of California?

What bay south-west of San José? (sahn ho-say'.)

Bound **OREGON**. What is its capital?

What mountains in the eastern part of Oregon? What range extends parallel with the coast?

What high mountain in the north?

What is the principal river in the north-western part of the United States?

What two large tributaries has the Columbia River?

What tributary of the Columbia drains the land between the Cascade Range and the coast?

What lake in the eastern part of Oregon? Southern?

Bound **WASHINGTON**. What is its capital?

What mountain range extends through Washington?

What large river flows into the Columbia?

What large island north-west?

What gulf east of Vancouver's Island? Strait south?

What sound projects into Washington?

What high mountain in the south?

What two capes on the coast?

Bound **IDAHO**. What is its capital?

On which side of Idaho are the Rocky Mountains?

By what rivers is Idaho drained?

What two rivers rise in the south-eastern part of Idaho and flow in different directions?

Bound **NEVADA**. What is its capital?

What mountains in the east? Lakes in the west?

What city near the capital?

Bound **MONTANA**. Name its principal cities.

What mountains in Montana?

What large river has its source in Montana on the west side of the mountains? On the east side of the mountains?

Bound **COLORADO**. What is its capital? What city near it?

What mountains extend through Colorado? Name the important peaks in Colorado.

What rivers drain the eastern slope? The western?

Bound **UTAH**. What is its capital?

What lakes in Utah? What rivers?

Bound **ARIZONA**. What is its capital?

By what rivers is it drained? What gulf receives their waters?

Bound **NEW MEXICO**. What is its capital?

What mountains in New Mexico? What rivers?

Bound **WYOMING**. What rivers flow through it?

Bound **ALASKA**.

Name the Territories. (*When a Territory becomes a State the pupils may erase from the Map the designation "Ty."*)

REVIEW.

CITIES AND TOWNS.

In what part of what States or Territories are they? On or near what waters? Which are capitals?

SAN FRANCISCO, 16	EUGENE CITY,	YANKTON,
	OLYMPIA,	SAN DIEGO,
SACRAMENTO, 18	PORTLAND,	IDAHO,
SALT LAKE CITY,	CARSON,	BANNOCK,
SALEM,		BOISEE,
DENVER,		PRESCOTT,
SAN JOSE,	SANTA FE,	GOLDEN CITY,
		VIRGINIA CITY.

MOUNTAINS.

Where are they? In what directions do the ranges extend?

ROCKY?	CASCADE?	SPANISH PEAK?
MT. ST. HELEN'S?	COAST?	PIKE'S PEAK?
	MT. SHASTA?	FREMONT'S PEAK?
SIERRA NEVADA?	SIERRA MADRE?	LONG'S PEAK?
		MT. HOOD?

RIVERS.

Where do they rise? What courses do they take? Into what waters do they flow?

MISSOURI?	KEHA PAHA?	OWYHEE?
COLORADO?		WHITE
GREEN?	GRAND?	YELLOW STONE?
BRAZOS?	NIOBRARA?	CLARKE'S?
SALMON?	PLATTE?	GILA?
COLUMBIA?	LEWIS?	RED?
SACRAMENTO?		WILLAMETTE?
HUMBOLDT?	CANADIAN?	SAN JOAQUIN?
RIO GRANDE?		

LAKES.

Where are they? What are their outlets?

GREAT SALT LAKE?	UTAH?	CARSON?
TULARE?	MUD?	KLAMATH?
SEVIER?	MALHEUR?	HUMBOLDT?
PYRAMID?	MINI WAKEN?	WALKER?

DESCRIPTIVE GEOGRAPHY.

1. *THE UNITED STATES:* This Republic comprises 38 States and 10 Territories, besides the District of Columbia, in which Washington, the capital, is situated.

2. *Its Extent* from east to west is about 2,800 miles, and from north to south, about 1,600 miles. Its area is over 3,000,000 square miles, not including Alaska, which covers over 400,000 square miles.

3. *The Largest State* is Texas, which is about half the size of Alaska, the largest Territory.

4. *The Greater Portion* of the United States is generally level or undulating. The high region comprises the western third of its area.

5. *The Great Mountain Chains* are the Rocky Mountains, the Sierra Nevada, and the Cascade Range.

6. *The Highest Peaks* are over 15,000 feet above the level of the sea.

7. *The Mountains in the Eastern part* are the Alleghany or Appalachian System, which extends from Northern Alabama and Georgia to Northern Maine. They comprise the Cumberland, Blue Ridge, Catskill, and Green Mountains. Their height is about one-fourth that of the Rocky Mountains.

8. *The Rocky and Alleghany Mountains Divide* the United States into three great physical regions—the Pacific Slope, between the Rocky Mountains and the Pacific Ocean; the Atlantic Slope, between the Alleghany Mountains and the Atlantic Ocean; and the Basin of the Mississippi River.

9. *Besides these* are the Gulf Slope, the St. Lawrence Basin (including the Great Lakes), and that part of the Red River Basin which is in Minnesota and Dakota.

10. *West of the Sierra Nevada* and the Cascade Range are fertile valleys in California, Oregon, and Washington Territory.

11. *A large portion of the High Region* is dry and barren, particularly the Great Basin in Nevada and Utah; but its river valleys are fertile and productive.

12. *The Pacific Slope* is rich, principally in its mineral products; the Mississippi Basin and Gulf Slope, in their agricultural products; and the Atlantic Slope, in its manufactures, commerce, agricultural and pastoral products.

13. *The Mountains, generally,* are covered with valuable timber, and nearly every State and Territory has its productive farms, pastures, and manufactories.

14. *This Country Possesses* nearly every variety of climate, soil, and productions. This is due to its great extent, its position on the globe, and the difference in elevation of the various parts of its surface.

15. *The Climate varies* according to the latitude, elevation, and the influences of the ocean and the mountain ranges.

16. *In the North-east* the winters are long and severe; the summers, hot and short.

17. *In the South* the summers are hot and the winters mild.

18. *Along the Pacific Coast* it is not so cold in winter, nor so hot in summer, as on the Atlantic coast, in corresponding latitudes (see page 37, paragraphs 21, 22, 23, 24, 25, and 38).

19. *In Minnesota and Westward* to the Rocky Mountains the winters are extremely cold, but remarkably dry and healthful.

20. *Rain* is abundantly supplied to the States between the Mississippi River and the Atlantic Ocean, by vapors from the Gulf and the Gulf Stream; and to those between the Sierra Nevada and the Pacific Coast, by vapors from the warm current of the Pacific Ocean; hence the fertility of these sections. Between them is the high region, where, in some places, rain but seldom falls, and the soil, consequently, yields little or no vegetation.

21. *The Desert Regions* of the United States are east of the Sierra Nevada, Cascade, and the Rocky Mountains, the west winds being deprived of their moisture before passing over the mountains (see page 34, paragraphs 41, 47, 48).

22. *The Agricultural Products* of the northern half of the United States are grain, fruits, and garden vegetables; of the southern half, cotton, tobacco, rice, and sugar.

23. *Celebrated for—*

Wheat, are Illinois, Iowa, Ohio, Indiana, & Wisconsin;
Indian Corn, Illinois, Iowa, Ohio, and Missouri;
Manufactures, the States N. of the Potomac;
Gold and Quicksilver, California;
Coal and Iron, Pennsylvania;
Silver, Nevada;
Copper, Michigan;

24. *Cotton,* the Gulf States, with N. and S. Car., Tenn. and Ark.;
Wool, California, Ohio, New York, Mich., and Penn.;
Tobacco, Kentucky, Virginia, and Tennessee;
Commerce, New York and Massachusetts;
Cane Sugar, Louisiana;
Rice, South Carolina;
Shipbuilding, Maine.

25. *This Republic Covers an area* more than nine times as large as that of the original thirteen States, which, previous to the 4th of July, 1776, were British colonies. Texas, with all that portion of the United States northwest to Oregon and the Pacific, was ceded to the United States by Mexico, in 1848. The remaining portion which lies between the Mississippi River and the Pacific Ocean, was ceded by France, in 1803; Florida was ceded by Spain, in 1819, and Alaska by Russia, in 1867.

26. *The Original Inhabitants* were Indians; the white people, who form the bulk of the population, are Europeans by birth or descent, and these are mainly of British or Irish extraction; next, are the Germans, French, and Swedes. The colored inhabitants are of African descent. The Chinese are numerous on the Pacific Slope.

27. *The First Settlements* were in Florida, in 1565; Virginia, in 1607; New York, in 1614; and Massachusetts, in 1620.

28. *The General Government* comprises the President and Congress. Congress is composed of a Senate and a House of Representatives.

29. *The President* is elected for four years.

30. *Each State is Entitled* to two Senators, who hold office six years. The number of its Representatives, whose term of office is two years, is according to its population.

31. *The whole number of Representatives* from the 37 States, in 1870, is 291; which, for a total population of 38,925,528, is one Representative for every 133,000 inhabitants. Each State is entitled to, at least, one Representative. Each Territory is allowed one, but he has no vote.

32. *The States having the largest number* of Representatives are: New York, 33; Pennsylvania, 27; Ohio, 20; Illinois, 19.

33. *Each State* is independent in the management of its internal affairs (see page 48).

34. *The United States* surpasses every other country in the world in its mineral and agricultural resources, in the extent of its rivers, canals, and railroads, in the enterprise of its people, in religious and political freedom and privileges, in its support of public instruction, and in the influence of the press.

1. ***THE NEW ENGLAND*** or North-eastern States occupy the north-eastern part of the Union. (See map on page 60.)

2. ***They are in the same Latitudes*** as Oregon and Southern France. (For their comparative climates, see page 37, paragraphs 21, 24, 31, 33, and 38.)

3. ***Their Characteristics*** are their rugged surface, their vast forests of pine, hemlock, spruce, etc.; their numerous streams and waterfalls, which furnish abundant water-power; and their long, severe winters.

4. ***Consequently,*** they are not well adapted to agriculture, but are celebrated for their manufactures, their lumber trade and shipbuilding, live stock, and dairy products. Their fisheries and coasting trade are also important. (See p. 37, par. 38.)

5. ***The Principal Mountains*** are the Green Mountains, so called from the evergreen forests which cover them, and the White Mountains, which are white with snow during most of the year.

6. ***The White Mountains*** are celebrated for their wild and picturesque scenery.

7. ***The Highest*** of the White Mountains are,—Mount Washington, over 6,000 feet, Mounts Jefferson, Adams, Madison, and Monroe, each over 5,000 feet above the level of the sea. Mount Katahdin, in Maine, is also over 5,000 feet high.

8. ***The Largest River*** is the Connecticut, whose valley is celebrated for its fertile soil and beautiful landscapes.

9. ***Maine Covers*** about one-half the area of New England, and excels every other State in the Union in the importance of its shipbuilding.

10. ***Its numerous Streams*** afford the means for floating its timber, and abundant water-power for sawing it into lumber (see page 30, paragraph 49).

11. ***Its Capital*** is Augusta, at the head of sloop navigation on the Kennebec River, and its metropolis is Portland, which is celebrated for its fine harbor and important railroad connections with Canada and the States.

12. ***A Village or a City is located,*** *usually with reference to some natural features;—on a bay or harbor, where ships may enter and anchor safely; on a river, for the purpose of navigation and trade; on a certain part of a river, as the head of navigation, or just below a waterfall or rapids where water-power for manufacturing purposes may be obtained; at or near the junction of two or more rivers; near mines or quarries; at the end of a lake, or at a mountain pass. In the old world many villages and towns were built in places almost inaccessible, that they might be easily defended against the attacks of enemies.*

13. ***The Surface*** of Maine, Vermont, and New Hampshire is more rugged and mountainous than that of Massachusetts, Connecticut, and Rhode Island.

14. ***The Manufacturing States*** are Massachusetts, Connecticut, New Hampshire, and Rhode Island.

15. ***Vermont*** is mainly a grazing and agricultural State.

16. ***The Capitals*** of New Hampshire and Vermont are Concord and Montpelier, and their chief cities, Manchester and Burlington.

17. ***Massachusetts Excels*** every other State in the Union in its manufactures of cotton and woolen goods, and in its whale and cod fisheries.

18. ***It is Second*** to Maine in shipbuilding, and to New York in commerce.

19. ***Its Capital and Metropolis*** is Boston, the largest city in New England; next in size in the State are Worcester, Lowell, Cambridge, Lawrence, and Charlestown. Boston is in nearly the same latitude as Detroit and Dubuque, in the United States, and as Rome, in Italy.

20. ***In Connecticut and Rhode Island*** the winters are shorter and less severe than in the other New England States.

21. ***The Principal Cities in Connecticut*** are New Haven, Hartford, Bridgeport, Norwich, Norwalk, and Waterbury. The principal cities in Rhode Island are its capitals, Providence and Newport.

22. ***Providence*** is the second city in size in New England. It is situated at the head of navigation on Narragansett Bay.

23. ***Newport*** has an excellent harbor, and is one of the most celebrated watering places in the United States.

24. ***The Legislature*** of Connecticut meets in Hartford and New Haven each year, alternately; of Rhode Island, in Providence, in winter, and in Newport, in summer.

25. ***The People*** of New England are chiefly of English descent. The first settlement was made at Plymouth, by the Pilgrim Fathers, December 22, 1620.

MAP DRAWING.

26. *To draw a map of a State, begin at its north-west corner, and proceed easterly, then southerly, then westerly, and northerly to the point of beginning.*

Then draw the mountains, the rivers, lakes, bays, capes, cities, and towns.

Draw only those rivers which are named on the map, and those cities and towns which appear in capital letters.

Where there is not space sufficient for the whole name, mark its first syllable or first letter. The New England States may be divided, and drawn on three different maps, being three separate lessons, viz.: 1st. Maine; 2d. New Hampshire and Vermont; 3d. Massachusetts; 4th. Connecticut and Rhode Island. (See Map Drawing in the Appendix.)

27. ***THE MIDDLE ATLANTIC STATES*** are New York, Virginia, and the intervening States. (See map on page 62.)

28. ***They Extend*** between 300 and 400 miles from the Atlantic coast.

29. ***They are all Mountainous*** except Delaware, and have two principal slopes; one eastward to the Atlantic, the other westward to the Ohio.

30. ***The Mountains*** are disposed in separate ranges, between which are fertile valleys. Their distance from the coast is greatest in Northern Georgia, about 250 miles; and least in New York and New Jersey, about 50 miles.

31. ***Eastward from the Mountains,*** and midway to the coast, is a hilly, fertile, and delightful country; while between the hill country and the coast it is generally low, with marshes and sandy islands along the coast.

32. ***Between the Hudson and the Potomac*** are States rich in agriculture, manufactures, and commerce; while south of the Potomac are those celebrated for the wealth of their agricultural products.

33. ***NEW YORK*** is in the same latitude as Oregon, Northern California, Northern Spain, and Southern France.

34. ***About one-half of its Surface*** is in the St. Lawrence Basin; the other half is drained by the Hudson, Delaware, Susquehanna and Alleghany Rivers, and the streams which flow into Lake Champlain.

35. ***Its Mountains*** and rich pastures are in the east; its level and agricultural lands, in the west.

36. *Near the Centre* are several lakes, celebrated for their beautiful scenery.

37. *In the North-east* is a wild, mountainous region, abounding in forests and lakes.

38. *In the South-east* are its two important islands, Long Island and Staten Island.

39. *The Highest Mountains* are the Adirondacks—their highest peak, Mt. Marcy, being over 5,000 feet high—and the Catskills, between 3,000 and 4,000 feet high.

40. *Between New York and Canada* are the Falls of Niagara, and the Thousand Islands of the St. Lawrence River.

41. *New York Excels* every other State in the Union in population, wealth, commerce, and importance of its canals and railroads.

42. *Its Capital* is Albany, and metropolis New York, the largest city on the western continent.

43. *New York City* is finely situated for commerce, having New York Bay on the south, the Hudson River on the west, and the East River on the east, all of which furnish wonderful facilities for shipping.

44. *Its Latitude* is the same as that of Naples and Constantinople (41° north latitude).

45. *Here took place* the first meeting of Congress, and the inauguration of Washington, the first President of the United States (1789).

46. *Brooklyn*, the second city in size in the State, is situated on the western extremity of Long Island.

47. *Buffalo* is situated at the eastern extremity of Lake Erie, the most southern of the Great Lakes. Its trade between the east and the north-west is extensive; owing chiefly to its position, fine harbor, important canal and railroad connections. Its manufactures are important.

48. *Rochester*, built on the Genesee River, at its falls, has long been noted for its immense flour mills. The city is crossed by the Genesee River and the Erie Canal, and a considerable trade is carried on with the east, west, and Canada.

49. *Troy* is situated on the Hudson River, at the head of steamboat navigation.

50. *Syracuse*, the most central of the large towns in New York, is noted for its manufacture of salt from the water of its salt springs or wells. The salt is obtained from the water by boiling or evaporation.

51. *Oswego* is the most populous and flourishing city on Lake Ontario (in New York). Its extensive commerce and manufactures are chiefly due to its fine harbor and its situation at the mouth and falls of the Oswego River.

52. *Niagara Falls, Syracuse, Utica and Saratoga Springs* are in nearly the same latitude as Milwaukee and Madison, in Wisconsin; and as Nice and Marseilles, in France.

53. *The First Settlements* were made where Albany and New York now stand, by the Dutch, in 1614. The latter place, then called New Amsterdam, received its present name in honor of the Duke of York, when it was taken by the English, in 1664.

Draw a map of New York, as directed on page 71.

54. *PENNSYLVANIA* is remarkable for its mountain ranges; the abundance of its coal and iron; its iron, cotton, and woolen manufactures; and its rich soil, which is well adapted to grain and grazing.

55. *Its Principal Range* extends through the center of the State; the highest peaks of this range—the Alleghanies—are between 2,000 and 3,000 fee high; of the Blue Ridge, in the south-eastern part, about 1,500 feet high.

56. *Its Capital* is Harrisburg; its metropolis is Philadelphia, which is the second city in population in the United States.

57. *Philadelphia* is a wealthy and important manufacturing city, situated on the west side of the Delaware River, and on both sides of the Schuylkill River, from which the city is supplied with fresh water. In its state-house the Declaration of Independence was signed by Congress, in 1776.

58. *Pittsburg* is situated at the confluence of the Monongahela and the Alleghany River, which here form the Ohio. It is surrounded by hills famous for coal and iron.

59. *Its Iron Works and Manufactures* are immense, and its coal trade, extensive.

60. *Among the other Important Cities* of Pennsylvania are Allegheny, Scranton, Reading, and Lancaster.

61. *Pittsburg is in the same Latitude* as Madrid, in Spain; Philadelphia, the same as Mount Ararat and Pekin, and in nearly the same latitude as Columbus, Indianapolis, Springfield, and Denver—four capitals in the United States.

62. *Pennsylvania was Settled* by the Swedes, in 1643. William Penn, after whom the State was named, established a colony here, in 1682.

Draw a map of Pennsylvania.

63. *NEW JERSEY*: Its northern half is hilly, with mountains in the north-west; its southern half, generally low and level.

64. *In the Central and Northern* portions are excellent farming and grazing lands and its flourishing manufacturing towns.

65. *The State is finely situated* for inland trade, owing chiefly to its position between the North-eastern and the Southern States. Its canals and railroads are of great importance.

66. *Its Capital* is Trenton, and metropolis, Newark: the other leading cities are Jersey City, Paterson, Elizabeth, Hoboken, Camden, and New Brunswick.

67. *Cape May*, *Long Branch*, and Atlantic City are famous summer and sea-bathing resorts.

68. *New Jersey* was settled by the Dutch, in 1620.

Draw a map of New Jersey, as directed on page 71.

69. *DELAWARE* is generally level—having hills only in the north.

70. *Its best Soil* is in the northern part.

71. *The Leading Agricultural Productions* are grain, garden vegetables and peaches.

72. *Its Capital* is Dover, and chief city, Wilmington, which has become celebrated for its various and important manufactures—particularly for its steamboat, car and iron works, its flour and powder mills.

73. *Wilmington is Situated* on the Christiana Creek, near its junction with the Brandywine.

74. *Delaware was Settled* by the Swedes and Finns, in 1638.

75. *MARYLAND* is level in the eastern, or widest part, and mountainous in the west, where it is narrow.

76. *It is rich* in coal, iron, tobacco, grain, cotton and commerce; its flour and cotton mills are extensive.

77. *The Capital* of Maryland is Annapolis.

78. *Its Principal Cities* are Baltimore, Frederick, and Cumberland.

79. *The First Settlements* were made by people from Virginia, in 1631, and from England, in 1634.

Draw a map of Maryland and Delaware, as directed on page 71.

80. ***VIRGINIA*** slopes eastwardly from the Blue Ridge, between which and the Alleghany Range is the fertile Valley of Virginia.

81. ***It is chiefly*** an agricultural State, and ranks next to Kentucky in the production of tobacco; its coal and iron mines and its abundant water-power furnish great facilities for manufactures.

82. ***Its Capital and Chief City*** is Richmond, situated at the lower falls of the James River and the head of navigation. It is remarkable for the beauty of its situation, and for its natural advantages as a manufacturing and business center.

83. ***Richmond is in the same Latitude*** as San Francisco and Mt. Etna.

84. ***Next in importance*** are Petersburg, Norfolk and Alexandria.

85. ***Among its Objects of Interest*** are numerous medicinal springs and the celebrated Natural Bridge, all in the western part of the State.

86. ***Virginia is noted*** as the birth-place of six Presidents of the United States,—Washington, Jefferson, Madison, Monroe, Tyler and Taylor, besides many other statesmen and officers closely identified with the independence and progress of the United States.

87. ***The First English Settlement*** in America was made on the James River, in 1607.

88. ***WEST VIRGINIA*** is a mountainous State, sloping northwestwardly from the Alleghanies to the Ohio River.

89. ***This State, like Virginia, contains*** rich mines of coal and iron, besides copper, lead and other metals.

90. ***It produces, also,*** grain, timber, tobacco, salt wool.

91. ***Its Capital*** and chief city is Wheeling, whose trade and manufactures are very important, owing mainly to its situation on the Ohio River, near extensive coal mines.

92. ***This State formed*** a part of Virginia until 1863.

Draw a map of Virginia and West Virginia, as directed on page 71.

1. ***THE SOUTHERN and SOUTH-WESTERN STATES*** are south of the parallel of 36° 30′ north latitude, with which the northern boundaries of three States—North Carolina, Tennessee and Arkansas—nearly coincide.

2. ***Bordering on the Atlantic*** are North Carolina, South Carolina, Georgia and Florida.

3. ***Bordering on the Gulf,*** are Florida, Alabama, Mississippi, Louisiana and Texas. These are called the Gulf States.

4. ***The Mountains*** are the southern portions of three ranges, here about midway between the Mississippi and the Atlantic. They are the Cumberland, Alleghany and Blue Ridge.

5. ***The Surface slopes*** from these mountains to the Atlantic, the Gulf, and the Mississippi River.

6. ***Along the Atlantic Coast*** is low, flat and marshy land, lined with islands which are celebrated for the production of Sea-island cotton.

7. ***The Interior*** is higher ground, with mountains in the west and north-west; except in Florida, Mississippi and Louisiana, which have no mountains.

8. ***Along the Lower Course*** of the Mississippi, the land is so low, that embankments have been raised from 5 to 10 feet high, to prevent inundation during the spring freshets; sometimes, however, the waters break over them and cause great destruction of property.

9. ***The Climate of the Low Lands*** along the coast and the Mississippi is unhealthy.

10. ***The Southern States possess*** a rich soil and a warm climate, which are peculiarly adapted to the production of cotton, rice, sugar and tobacco.

11. ***NORTH CAROLINA: The Eastern part*** is low, sandy and marshy, and is noted for turpentine forests, palmetto groves, and the production of rice, cotton and tobacco.

12. ***The Western part*** is mountainous and a good grazing country; portions are moderately high, and well adapted to the raising of grain.

13. ***Its Capital*** is Raleigh, and its chief city and seaport is Wilmington.

14. ***The First Permanent Settlement*** was made by colonists from Virginia, in 1653. Sir Walter Raleigh made an unsuccessful attempt to settle the State, in 1586.

Draw a map of North Carolina, as directed on page 71.

15. ***SOUTH CAROLINA and GEORGIA*** have the same general characteristics as North Carolina, only warmer and less mountainous; the northern half of each being high and adapted to the growth of grain; the southern half is low, producing rice and cotton.

16. ***Their Capitals*** are Columbia and Atlanta, and their chief cities Charleston and Savannah, both of which have excellent harbors.

Draw a map of South Carolina.

17. ***FLORIDA:*** its characteristics are its low, marshy surface, its warm, moist and even climate, its forests and luxuriant vegetation. Oranges and other tropical fruits are cultivated.

18. ***Its Capital*** is Tallahassee.

19. ***The First Settlement*** was made by the Spaniards, in 1565, at St. Augustine, which is the oldest town in the United States.

Draw a map of Florida.

20. ***ALABAMA*** has a general slope to the Gulf, with mountains in the north, hills in the center, and low lands in the south.

21. ***Its Northern part*** is drained by the Tennessee River, which enters the State at its north-eastern corner, and leaves it at its north-western.

22. ***The State contains*** forests of pine and cypress, and yields extensively, cotton, grain, sugar and sweet potatoes.

23. ***Its Capital*** is Montgomery, and its chief city, Mobile; the latter is celebrated as a shipping port for cotton.

24. ***Alabama was Settled*** by the French, in 1702, and admitted into the Union, in 1819.

Draw a map of Alabama.

25. ***MISSISSIPPI*** and Alabama have less sea-coast than the other Gulf States.

26. ***Mississippi*** is low and level in the west and south. It is noted for its production of cotton, rice and tobacco.

27. ***Its Capital*** is Jackson; its largest towns are Vicksburg and Natchez.

28. ***It was Settled*** by the French, in 1716, and admitted into the Union, in 1817.

Draw a map of Mississippi.

29. ***LOUISIANA*** has a low surface, which in some places is below high water level.

30. ***Along the Mississippi,*** on both sides, are extensive marshes.

31. ***Louisiana excels*** in the production of cane sugar; besides this, are cotton, corn, rice and tropical fruits.

32. ***New Orleans*** is the capital and metropolis of the State.

33. ***It is celebrated*** for its immense trade, especially in cotton.

34. ***It is built*** around a bend in the river, and is therefore sometimes called the "Crescent City."

35. ***Its Surface*** is below the level of the Mississippi during the usual freshets, but the city is well protected from inundation by its levee, which is about six feet in height and of a considerable width.

36. ***The Latitude*** of New Orleans is the same as that of Cairo, the capital of Egypt (30°).

37. ***Among its other Important Cities*** are Shreveport and Baton Rouge. The latter, until lately, was the capital of the State.

Draw a map of Louisiana.

38. ***TENNESSEE,*** a south-western State, is divided by the Cumberland Mountains and the Tennessee River into East Tennessee, which is mountainous; Middle Tennessee, hilly; and West Tennessee, generally level.

39. ***The Soil*** is fertile, and the climate delightful.

40. ***Its Productions*** consist chiefly of live stock, corn, cotton, and tobacco.

41. ***Its Trade*** with other States is extensive.

42. ***The Abundance of Iron, Coal, and Water-power*** in East Tennessee furnishes great facilities for manufacturing purposes.

43. ***Its Capital*** is Nashville, situated at the head of steamboat navigation on the Cumberland River. It is a beautiful and flourishing city, and is next in size to Memphis, the metropolis of the State.

44. ***Nashville is in the same Latitude*** as Gibraltar, Algiers and the southern part of Greece.

45. ***Tennessee*** originally formed part of the possessions of North Carolina, and became a State in the Union, in 1796.

Draw a map of Tennessee.

46. ***ARKANSAS*** has a south-easterly slope from its mountains in the north-west, to its extensive marshes along the Mississippi.

47. ***Its Leading Productions*** are corn, cotton and live stock.

48. ***Its Forests*** of valuable timber and its prairies are extensive.

49. ***Little Rock*** is its capital and chief city.

50. ***Its Minerals***—coal, iron, lead and zinc—are abundant.

Draw a map of Arkansas.

1. ***All the States North and North-west of the Ohio River*** are remarkable for their rapid increase in wealth, population and internal improvements. (*See Map on page 66.*)

2. ***Their Surface***, except that of Minnesota, is chiefly prairie land, level or rolling, which possesses wonderful fertility, especially along the rivers.

3. ***The Prairie States*** are Ohio, Indiana, Illinois, Wisconsin, the southern peninsula of Michigan, Iowa, Missouri, Arkansas, Kansas and Nebraska.

4. ***There are no Mountains*** except in Southern Missouri and North-western Arkansas.

5. ***They are Drained*** by the Mississippi or its tributaries, except Michigan, which is in the Basin of the St. Lawrence or the Great Lakes.

6. ***The rapid Progress*** of these States is due chiefly to their fertile soil, the facilities for manufactures and commerce furnished by their mines, rivers, lakes, canals and railroads, and to the energy of their inhabitants.

7. ***Their Agricultural and Grazing Products*** are immense—grain, live stock and wool.

8. ***Their Mines*** are of coal, iron, lead, copper and zinc.

9. ***Coal and Iron*** abound in nearly all, especially in Ohio and Missouri; lead, in Wisconsin, Iowa, Illinois, and Missouri.

10. ***OHIO*** has two general slopes; the larger is southerly to the Ohio River, and the smaller, northerly to Lake Erie.

11. ***It is between the same Parallels*** of latitude as northern California and southern Italy.

12. ***Its Commerce*** is extensive, having outlets, northerly, by way of Lake Erie, Welland Canal, Lake Ontario and the St. Lawrence River, and southerly, by way of the Ohio and Mississippi Rivers; its railways are numerous and important.

13. ***The Coal, Iron and Petroleum Region*** is in the south-east.

14. ***Its Capital*** is Columbus; its chief city, Cincinnati, on the Ohio, opposite the mouth of the Licking, a river of Kentucky.

15. ***The Second City*** in size is Cleveland, on Lake Erie, at the mouth of Cuyahoga (*ki-a-ho'ga*) River.

16. ***Cincinnati and Cleveland*** are wealthy commercial and manufacturing cities, although at the beginning of the present century, neither contained 800 inhabitants.

17. ***The other Leading Cities*** are Toledo, Dayton, and Sandusky.

18. ***Ohio,*** with the other States west and north-west to the Mississippi River, was ceded by France to Great Britain, in 1763, and at the Revolution, came into the possession of the United States Government, known afterwards as the North-west Territory.

Draw a map of Ohio.

19. ***INDIANA and ILLINOIS*** have the same general characteristics as Ohio; each having soil of remarkable fertility, and facilities for commerce by lake, river, rail and canal.

20. ***Their Surface Slopes*** south-westerly.

21. ***Their yield of Wheat, Corn and Wool*** is immense.

22. ***Indianapolis,*** the capital and metropolis of Indiana, and an important railroad centre, is situated near the middle of the State.

23. ***The Cities next in rank*** are Evansville, Fort Wayne, and Terre Haute.

24. ***The Largest Town*** in the northern part of Indiana is Fort Wayne, an important railroad center.

Draw a map of Indiana, as directed on page 71.

25. ***ILLINOIS:*** its principal river, the Illinois, has its headwaters within but a few miles of Lake Michigan.

26. ***Its Chief City*** is Chicago. Its site, although but little above the level of Lake Michigan, is 600 feet above the ocean.

27. ***Illinois excels*** in the production of corn, wheat, and oats.

28. ***Chicago has excelled*** every other city in the world in the rapidity of its growth and development. In 1831, it contained about a dozen families, beside the officers and soldiers in Fort Dearborn. It is now the fifth city in size in the Union.

29. ***It excels every other City*** in the United States, as a market for grain, lumber, beef and pork.

30. ***The other large Cities in Illinois*** are Quincy and Peoria, which are nearly in a line south-west of Chicago.

31. ***Its Capital*** is Springfield, which is in a line between Chicago, Alton and St. Louis.

32. ***Galena*** is situated in the lead region of Illinois.

Draw a map of Illinois.

33. ***MICHIGAN***, the "Lake State," comprises two peninsulas.

34. ***The Southern and larger Peninsula*** contains rich agricultural and grazing land; grain, live-stock and wool being largely produced.

35. ***The Watershed*** which divides the eastern from the western slope, is about 300 feet above the level of the lake, and 1,000 feet above that of the sea.

36. ***The Northern Peninsula*** is noted for its rugged and mountainous surface, rigorous climate, and its rich mines of copper and iron.

37. ***Both Peninsulas*** contain extensive forests of valuable timber.

38. ***The Trade*** of Michigan by lake and rail is very extensive.

39. ***The Southern Peninsula*** lies between the same parallels of latitude as Oregon and the northern half of Italy.

40. ***The Capital*** is Lansing, and its chief city Detroit, which is finely situated for commerce, manufactures, lumber trade and ship building.

41. ***The other Important Cities*** are Grand Rapids, Jackson, and East Saginaw.

Draw a map of Michigan.

42. ***WISCONSIN*** resembles Michigan in its prairies, forests, trade, and its agricultural, grazing and mineral products.

43. ***Lead*** is abundant in the south-western part of the State.

44. ***Its Surface slopes*** mainly south-westerly to the Mississippi, and contains numerous rivers and lakes.

45. ***Madison***, the capital, is beautifully situated between two lakes in the southern part of the State.

46. ***Milwaukee***, the metropolis, is remarkable for its fine harbor, on Lake Michigan at the mouth of the Milwaukee River, its rapid growth, immense wheat trade, its flour and other manufactures.

47. ***Its other Leading Cities*** are Fond du Lac, Oshkosh, and Racine.

48. ***At Portage City*** is a canal connecting the Wisconsin and Fox Rivers, thus establishing water communication between the Mississippi River and the Great Lakes.

Draw a map of Wisconsin.

49. ***KENTUCKY***: its surface slopes from the Cumberland Mountains on the east, to the Ohio River.

50. ***Its Soil*** is fertile. Its leading productions are tobacco, corn, wheat, and hemp; the raising of live-stock is important.

51. ***Kentucky*** is remarkable for its caverns, the most wonderful being the "Mammoth Cave," south of the middle of the State.

52. ***Frankfort*** is the capital, and Louisville, the chief city. Covington, Newport, and Lexington are important cities.

53. ***Kentucky***, which formerly belonged to Virginia, was explored in 1770, by Daniel Boone, a hunter from North Carolina. (He was born in Pennsylvania.)

Draw a map of Kentucky.

54. ***MISSOURI*** is generally level, with a gentle slope eastward to the Mississippi River.

55. ***In the South and South-west*** is a ridge of mountains (the Ozark), and in the south-east, an extensive marsh.

56. ***Its Soil*** is productive, especially along the rivers.

57. ***Its Leading*** agricultural and grazing products are grain, hemp and live-stock.

58. ***It is Rich*** in iron, lead, coal and other minerals; noted for iron, are Iron Mountain, Pilot Knob and their vicinity.

59. ***Its Capital*** is Jefferson City; its chief city, St. Louis, a great commercial and manufacturing center. It is the largest city on the Mississippi River, except New Orleans.

60. ***Kansas*** is the second city in size in the State, and the largest on the Missouri River. St. Joseph also is a flourishing city.

Draw a map of Missouri.

61. ***IOWA***: its surface is highest in the north-west, and slopes south-easterly to the Mississippi. It is chiefly prairie land, with trees along the river banks.

62. ***It is Rich*** in agricultural and grazing products—grain, potatoes, live-stock, wool, etc., and in mineral products—lead, coal, iron, copper, and zinc.

63. ***Its Capital*** is Des Moines, which is situated at the head of steam navigation on the Des Moines River. It possesses extensive water power.

64. ***Its Largest Cities*** are situated on the Mississippi River; they are Davenport, Dubuque, Burlington, and Keokuk, all of which possess facilities for manufactures and inland trade.

65. ***Iowa was Admitted*** into the Union, in 1846.

Draw a map of Iowa.

66. ***MINNESOTA*** is the most northern State in the Union, and one of eight States which border on one or more of the Great Lakes.

67. ***Its Elevation*** is about 2,000 feet above the level of the sea, and its highest land is a watershed from which rivers flow into the Gulf of Mexico, Hudson's Bay and Lake Superior.

68. ***The Sources*** of the Mississippi and the Red River of the North are in this State.

69. ***Its Leading Characteristics*** are its extensive forests and prairies, its numerous lakes and streams, its rapids and waterfalls, and its cold winter climate, which is remarkably dry, even and healthful.

70. ***The Leading Occupations*** of the inhabitants are agriculture and the lumber trade.

71. ***The North-eastern part*** of the State is a rich mineral region.

72. ***The Leading Cities*** are St. Paul, the capital, at the head of steamboat navigation, Minneapolis and Winona; all are situated on the Mississippi River, and possess extensive steam saw-mills and flouring mills.

73. ***St. Paul is in the same Latitude*** as Salem, in Oregon, and Bordeaux, Turin and Sevasto'pol, in Europe.

74. ***Minnesota formerly belonged*** to France; the portion east of the Mississippi was ceded to Great Britain, but came into the possession of the United States, at the Revolution; the portion west of the river was ceded to the United States by France (see page 70, paragraph 25).

Draw a map of Minnesota.

1. ***TEXAS***, the largest State in the Union, is as large as France, Holland and Belgium combined.

2. ***It is between the same Parallels*** of latitude as Egypt and Morocco.

3. ***Its Characteristics*** are its rolling prairies, where vast herds of horses and cattle graze throughout the year; its desert plateau in the west and northwest, and its low land in the south.

4. ***Its Slope*** is southerly and south-easterly to the Gulf.

5. ***Its Climate*** is warm and healthful; ice or snow being seldom seen in some parts; that of its southerly half is tropical and temperate.

6. *Texas is Rich* in grass, live-stock, cotton, sugar, corn, wheat, etc.

7. *Eastern Texas* is inhabited, principally, by people from other southern States, while in Western Texas are large numbers of Germans and other Europeans.

8. *Wild Animals* are numerous,—buffaloes, mustangs (wild horses), deer, bears, wolves, etc.

9. *Its Capital* is Austin, at the head of steamboat navigation on the Colorado River; its largest cities are Galveston, San Antonio, and Houston. Galveston, on Galveston Island, is noted for its fine bay and its extensive commerce.

Draw a map of Texas.

10. *KANSAS and NEBRASKA* correspond in latitude with Spain and Portugal.

11. *Their rich Rolling Prairies* and fine climate make them very productive agricultural and grazing States.

12. *Coal and Salt* are found in abundance.

13. *Their Increase* in population and general importance has been very rapid.

14. *Kansas* is the most central State in the Union; its capital is Topeka, and its chief cities are Leavenworth, Lawrence, and Atchison, all in the eastern part of the State.

15. *The Capital of Nebraska* is Lincoln, and its chief cities are Omaha and Nebraska City.

16. *The Union Pacific Railroad* passes through Nebraska, and the Kansas Pacific through Kansas.

Draw a map of Kansas and Nebraska.

17. *INDIAN TERRITORY* is occupied by several tribes of Indians.

18. *Its Soil* is well adapted to agriculture and grazing.

19. *DAKOTA* is but thinly inhabited. Like the other Territories, the development of its resources is retarded by Indian disturbances. It is crossed by the Northern Pacific Railroad.

20. *ARIZONA and NEW MEXICO* are noted for their high mountains and plateaus, deep cañons and gorges, and their mineral resources.

21. *The Climate* is dry and healthful; and the soil of the river valleys, well adapted to grazing and agriculture.

22. *The Eastern part of New Mexico* forms part of the high desert region extending from northern Texas into eastern Colorado.

23. *COLORADO* lies on both sides of the Rocky Mountains, between which are several beautiful plains called "Parks." Some of the peaks are over 13,000 feet above the level of the sea.

24. *It is noted* for its wealth in precious metals; it is also rich in extensive tracts of fertile soil, producing grain, fruits and grasses in abundance.

25. *Its Climate* is salubrious.

26. *UTAH* is elevated, mountainous and barren.

27. *Its Western Section* forms part of the "Great Basin," which is over 4,000 feet above the level of the sea. Its soil contains salt and is generally dry and unproductive.

28. *The principal Valley*, in which is situated the capital, Salt Lake City, is southeast of Great Salt Lake.

29. *Utah is noted* as the residence of the Mormons.

30. *WYOMING*, like Colorado, lies partly on the Atlantic and partly on the Pacific Slope. It is high, mountainous, and but little settled. The Union Pacific Railroad passes through it.

31. *CALIFORNIA* is the largest State except Texas. Its northern boundary corresponds in latitude with that of Pennsylvania, and its southern with that of South Carolina.

32. *Its principal Mountain Ranges* are the Sierra Nevada and the Coast Range, which enclose the great valley of California, noted for its fertility.

33. *The Sierra Nevada* rise generally above the snow limit, having many peaks from 7,000 to 15,000 feet in height.

34. *The Yosemite (yo-sem'-i-te) Valley*, in the Sierra Nevada, is celebrated for the grandeur of its scenery. Its length is 8 miles, and its width is less than one mile. It is enclosed by mountains which rise almost perpendicularly more than 4,000 feet.

35. *The Climate* of California, Oregon and Washington, is milder than that of the other States of the Union in the same latitude. (See page 37, paragraphs 21 to 38.)

36. *California is celebrated* for its mineral and agricultural productions. Its commerce and manufactures are rapidly increasing in importance.

37. *The leading Mineral Products* are gold, quicksilver and silver.

38. *The Forests* of California, and the western sections of Oregon and Washington, abound in magnificent pines and a variety of trees valuable for timber. The "Big Trees" of California, are 350 feet high and 30 feet in diameter.

39. *San Francisco* is the largest city west of the Rocky Mountains. Next in size is Sacramento, the capital.

40. *NEVADA* is chiefly a vast basin at an elevation of over 4,000 feet above the sea level.

41. *Its Mountain Ranges* are short and numerous. The rivers are small, and empty into lakes, which, having no outlets, are generally salt or alkaline.

42. *The Climate* is dry and the soil barren, but the State possesses great wealth in its silver mines.

43. *The Leading City* is Virginia.

44. *OREGON* is situated on the Pacific Coast, in a line directly west of New England.

45. *Its area* is equal to that of New York and Pennsylvania combined.

46. *The State is divided* by the Blue and the Cascade Mountains into three physical sections,—the Western, Middle and Eastern, styled respectively the Lower, Middle, and Upper Counties.

47. *The Rich Valleys*, and most of the inhabitants of Oregon and Washington, are between the Cascade Mountains and the Pacific Ocean. Grain, flour and wool, are exported from Oregon; lumber and ship timber from both.

48. *Portland* is the principal city in Oregon, and the most important city on the Pacific Slope north of California.

49. *WASHINGTON TERRITORY* is in the same latitude as the northern part of Maine and the central part of France.

50. *It is divided* by the Cascade Range into two physical regions,—the western or lower, and the eastern or higher.

51. *Olympia* is the capital.

52. *IDAHO and MONTANA* are noted for gold and silver. Their surface is mountainous, but well adapted to grazing. The Northern Pacific Railroad passes through those Territories.

53. *The Chief Towns* in Idaho are Boisé City and Lewiston, and in Montana, Helena and Virginia City.

CHART SHOWING THE COMPARATIVE AREAS OF STATES, COUNTRIES, Etc.

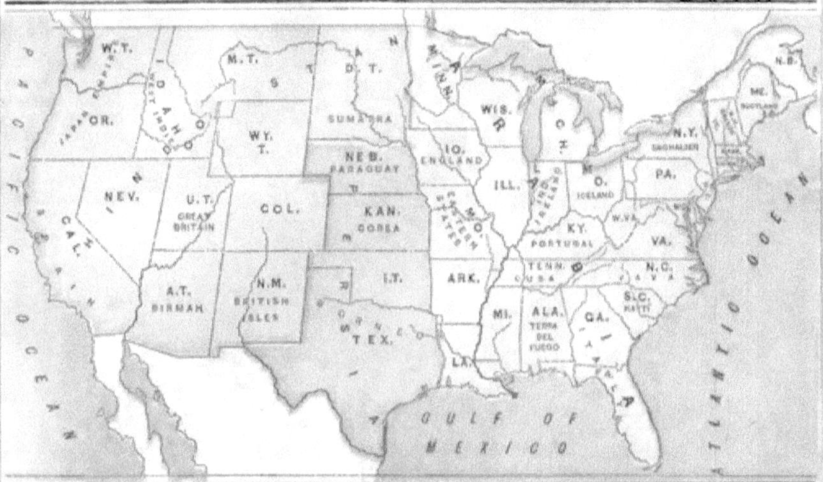

The Countries and Islands have areas equal, or nearly equal, to those of the States in which their names appear. For example, the area of Iowa is about the same as that of England.
The portion of the map which is colored yellow represents the area of Arabia; that colored red, Hindoostan; green, Persia.

What three countries in Asia taken together cover an area nearly equal to that of the United States, without Alaska?

How many States together comprise an area equal to that of Arabia? *Ans.* 31.

What States are included in this area? *Ans. All those between the Atlantic Ocean and the Mississippi River, besides the five States which are situated on the west side of that river.*

What State is in the north-eastern part of the region which is here compared with Arabia? In the south-western part? North-western? South-eastern part?

What rivers flow through that part of the United States?

What States and Territories together comprise an area equal to that of Hindoostan?

What mountains extend through that part of the United States?

What are included in the region whose area is equal to that of Persia?

Bound that part of the United States which has the same area as Arabia. Persia. Hindoostan.

Which is the largest State in the Union?

What country in Europe has about the same area as California? Kentucky? Maine? Iowa? Indiana?

What country is the same size as New Hampshire and Vermont combined? What country is but little larger than Georgia and Florida combined?

What country in South America is nearly as large as Nebraska?

What empire has an area nearly equal to that of Oregon and Washington combined?

What islands comprise an area equal to that of Idaho? Of New Mexico?

What island has an area nearly the same as that of the State of New York? Ohio? Alabama? North Carolina? Dakota? Tennessee? South Carolina? Utah? Texas?

What six States have the same area as Missouri?

What country in Asia contains the same number of square miles as Kansas?

What country in Asia is the same size as Arizona?

What State or what Territory has an area equal to that of

Great Britain?	Scotland?	Corea?	Saghalien?
Spain?	Ireland?	Paraguay?	The Eastern or New England States?
Portugal?	Iceland?	Terra del Fuego?	West Indies?
The British Isles?	Birmah?	Sumatra?	Cuba?
England?	Hayti?	Java?	

What part of the United States has the same area as Italy? Greece? Japan Empire? Borneo?

AREAS IN SQUARE MILES.

	Sq. miles.		Sq. miles.		Sq. miles.
Maine	35,000	Tennessee	45,600	Missouri	65,350
Scotland	31,324	Cuba	43,000	Eastern States	66,000
Vermont & N. Hamp.	19,000	N. Carolina	50,704	California	189,000
Greece & Ionian Is.	19,000	Java	51,000	Spain	195,000
Ohio	39,964	S. Carolina	34,000	Montana	143,000
Iceland	39,000	Liberia	95,000	Prussia	135,000
Indiana	33,809	Georgia & Florida	109,000	Hindoostan	1,500,000
Ireland	32,509	Italy	114,000	Arabia	1,000,000
				Persia	500,000
England	50,922	Texas	274,356	Hind., Ara. & Persia	3,000,000
Iowa	55,000	Borneo	284,000	Australia	3,000,000
Kentucky	37,680	New Mexico	121,000	United States, without Alaska	3,000,000
Portugal	36,000	British Isles	120,000		

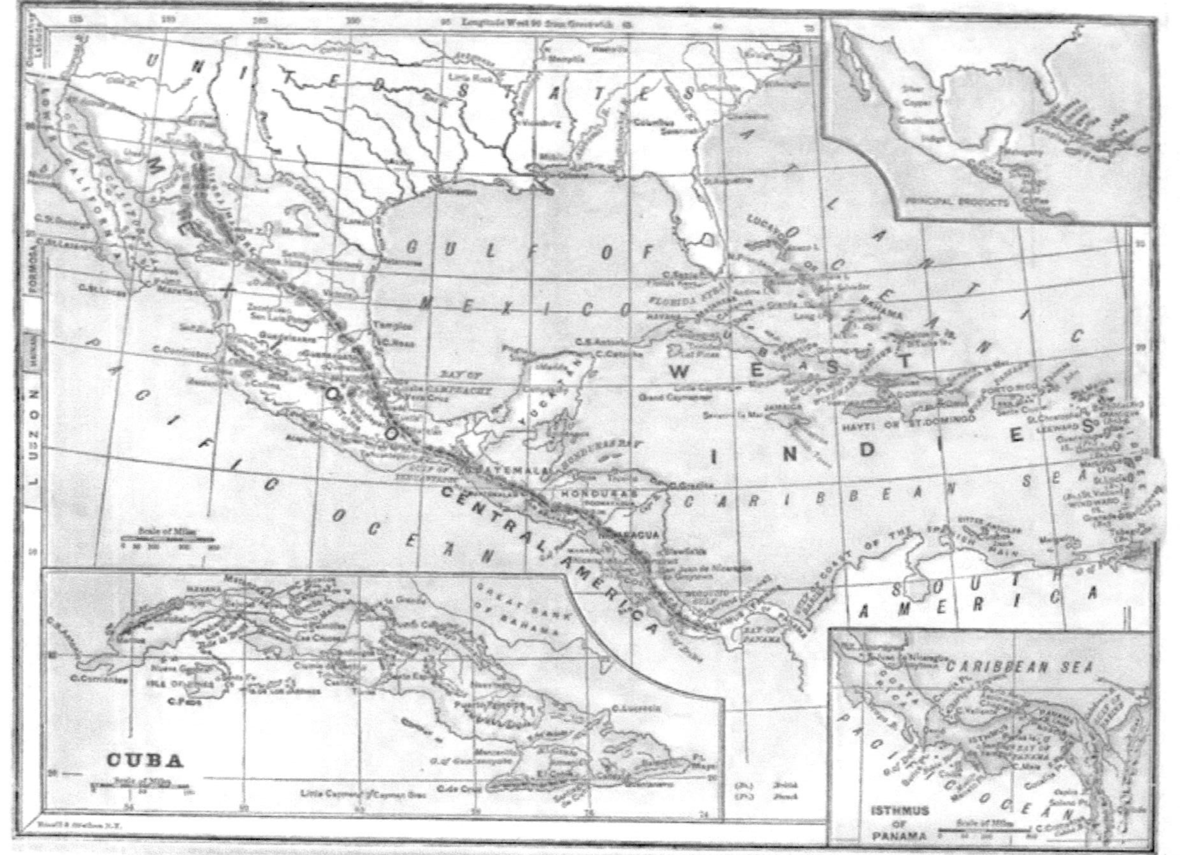

UNITED STATES
MEXICO
GULF OF MEXICO
ATLANTIC OCEAN
WEST INDIES
CARIBBEAN SEA
PACIFIC OCEAN
CENTRAL AMERICA
SOUTH AMERICA
GUATEMALA
HONDURAS
NICARAGUA
BAY OF CAMPEACHY
HAYTI OR ST. DOMINGO
JAMAICA
BAHAMA
LUZON
FORMOSA
HAINAN
LOWER CALIFORNIA
PRINCIPAL PRODUCTS
CUBA
GREAT BANK OF BAHAMA
ISTHMUS OF PANAMA
CARIBBEAN SEA
PACIFIC OCEAN

SECTION OF MEXICO FROM THE PACIFIC TO THE GULF OF MEXICO.

EXERCISES ON THE MAP.

Bound **MEXICO.** What country north and north-east?

On the south-east what state of Central America and what British Territory?

What great mountain range in Mexico? In what direction does the Sierra Madre (*se-er'rah mah'dray*) range extend?

In what direction does Mexico extend?

What peninsula in the south-east? In what cape does Yucatan terminate? By what waters is it embraced?

What isthmus in the south-east? What waters are separated by it?

What peninsula in the north-west? By what waters is Lower California embraced?

In what direction does it extend? In what cape does it terminate?

By what gulf and river is Lower California separated from Mexico?

In what direction do the rivers of the western part of Mexico flow?

In what direction do those of the eastern part flow?

At what part of Mexico do the waters of the Gulf of Mexico approach nearest to those of the Pacific Ocean?

What two cities on the eastern coast? What one on the western coast?

What cities in the interior? Which one is between Vera Cruz and Mexico?

What active volcano near the City of Mexico?

What cape on the eastern coast? What capes on the western coast?

Bound **CENTRAL AMERICA.** In what direction does its mountain chain extend?

In what direction does Central America extend?

In what direction does its western coast extend?

In what direction does its eastern coast extend?

Which of the five divisions of the earth does Central America resemble in form?

Name the divisions or states of Central America.

Which is farthest north? South? Which is the smallest?

On which coast is San Salvador? On which is the British territory of Balize? (*bah-leez'.*)

What lake affords the nearest approach of boats from the Atlantic to the Pacific side of the Continent?

What bay north of Honduras?

What gulf east of Costa Rica? (*kos'tah re'kah.*)

What is the capital of Guatemala? Honduras? San Salvador? Nicaragua? Costa Rica?

How many islands do the West Indies comprise? Ans. About 1000.

Near which of the United States are the most northern of the West Indies?

Near what coast are the most southern of these islands?

What group of the West Indies farthest north? Last?

What island is farthest east? South-east?

What gulf and delta near Trinidad Island?

Mention the largest four Islands of the West Indies?

By what name are they known? Ans. The Great Antilles. (*an-teel'.*)

What group off the northern coast of South America?

In what direction do the Great Antilles extend? The Little Antilles?

In what direction do the Caribbee Islands extend?

Bound **CUBA.** What is its capital? To what government does Cuba belong? (*See Review.*)

What maritime town east of Havana? What town near the center?

Is Havana north, or south, of the Tropic of Cancer?

Which of the Bahama Islands are north of the Tropic of Cancer? What island is covered by it?

What keys or reef south of Florida? What is here meant by keys? Ans. A ledge of rocks at or near the surface of the water.

What cape at the western extremity of Cuba? What islands south?

Bound **HAYTI.** Of what two republics is the island of Hayti (*hay'te*) composed?

What part of the island is comprised in the Republic of Hayti? In the Republic of Dominica?

What is the capital of the Republic of Hayti? Of Dominica?

What island east of Hayti? What is the capital of Porto Rico? (*por'to re'ko.*)

Bound the **CARIBBEAN SEA.** What two gulfs south? What bay west?

Bound the **GULF OF MEXICO.** What bay south?

ROUTES OF TRAVEL.

In what directions and on what waters would you sail from Baltimore to Vera Cruz? New Orleans to Vera Cruz? New Orleans to the Isthmus of Panama?

What towns on opposite sides of the isthmus connected by railroad?

On what waters would you sail from the mouth of the Orinoco to that of the Rio Grande? From Panama to Mazatlan?

On what waters would you sail and what capes would you pass from Tampico to Costa Rica?

What is the distance between New Orleans and Havana? Washington and Aspinwall? Havana to Savannah? Cuba to Florida? Cuba to Yucatan?

What is the length of Cuba? Of the Gulf of Mexico?

REVIEW.

COUNTRIES.

Where are they? What sea coast have they?

Mexico?	Nicaragua?	Guatemala?	Costa Rica?
Honduras?	(*[illegible]*)	(*[illegible]*)	Balize?

CAPES.

Where are they? Into what waters do they project?

St. Antonio?	Gracios?	Palmo?	Roxo?
Corrientes?	Burica?	St. Lucas?	Cruz?

GULFS AND BAYS.

Where are they? Into what waters do they open?

California?	Tehuantepec?	Honduras?	Panama?
Campeachy?	(*[illegible]*)	Mosquito?	Darien?

STRAITS.

Between what lands are they? What waters do they connect?

Florida?	Yucatan?	Windward?	Mona?

ISLANDS.

Where are they? By what waters surrounded? To what nations do they belong?

To Great Britain.

Jamaica?	New Providence?	Andros?
Bahamas?	San Salvador?	Aba'co?
Barba'does?	Antigua? (*[illegible]*)	St. Vincent?
Trinidad?		Turks?

To France.	To Spain.	To Denmark.
Guadaloupe?	Cuba?	St. Thomas?
Martinique? (*[illegible]*)	Porto Rico?	Santa Cruz? (*[illegible]*)
	I. of Pines?	

CITIES AND TOWNS.

Where are they? On or near what waters? Which are capitals?

MEXICO.	GUATEMALA.	VERA CRUZ.
HAVANA.	SPANISH TOWN.	MAZATLAN.
MATANZAS.	ST. DOMINGO.	PANAMA.
PORT AU PRINCE.	SAN SALVADOR.	BALIZE.
GUADALAXARA. (*[illegible]*)	SAN JUAN. (*[illegible]*)	SAN JOSE. (*[illegible]*)
GUANAXUATO. (*[illegible]*)	TAMPICO. (*[illegible]*)	PUEBLA. (*[illegible]*)
PUERTO PRINCIPE. (*[illegible]*)	COMAYAGUA. (*[illegible]*)	MANAGUA. (*[illegible]*)

ARGENTIN
EPUBLI
Santa Fe
OR
San Luis
LA PLATA

SECTION OF SOUTH AMERICA THROUGH THE CENTER FROM THE PACIFIC TO THE ATLANTIC.

EXERCISES ON THE MAP.

What division of land is SOUTH AMERICA? In what direction does it project?

To what is it joined? By what isthmus?

What water north of South America? East? West?

What cape at its northern extremity? Eastern? Southern? Western?

Between what capes is South America longest?

If you draw a line between these two capes, on which side of the line would be the greater part of South America?

Between what two capes is the greatest width of South America?

Is the greater part of South America north or south of its widest part?

On which side is its great mountain chain? With what coast is that chain parallel?

What mountains in its eastern part? With what coast are they parallel?

What mountains on the northern boundary of Brazil?

What mountains in the central part of South America?

Into what ocean do nearly all the rivers flow?

What river of South America is the largest in the world?

Is the Amazon north or south of the Equator?

What line on the map passes through its mouth?

What tributaries flow into its southern side? Northern side?

Which have their sources in the Andes?

What two countries of South America wholly north of the Equator?

What three countries partly north of it?

What country between the Andes Mountains and the Pacific Ocean?

Name all the countries which border on the Pacific? On the Atlantic? On the Caribbean Sea?

What is the only country without sea coast?

In what country is the Isthmus of Panama included?

Bound the UNITED STATES OF COLOMBIA. What is the capital?

What cities in the north? What city in the south-west?

What two rivers flow north? South-east?

Bound VENEZUELA. What is its capital?

What city and lake in the north-west?

What large river flows through Venezuela?

What island north of the delta of the Orinoco? What gulf?

Bound GUIANA. Into what three colonies is it divided?

Which colony is in the eastern part? Western? Central?

What is the capital of each colony?

Bound BRAZIL. What is its capital?

Name the capes on the coast of Brazil. The cities.

What large island at the mouth of the Amazon?

In what part of Brazil are its mountain regions? Its Silvas? Llanos? Pampas? Its diamond district?

What lake in the southern part of Brazil?

By what river and its tributaries is the greater part of Brazil drained?

By what is its southern part drained? Its eastern?

What is the greatest width of Brazil from east to west? Of the United States of North America?

Bound ECUADOR. What is its capital?

What three celebrated volcanoes near Quito? What high mountain?

What two capes on the coast of Ecuador?

What tributaries of the Amazon flow through Ecuador?

Bound PERU. What is its capital? What is the port of Lima?

What town in the north-west? What town and volcano in the south?

What capes on the coast? What lake and river in Peru?

Bound BOLIVIA. What is its capital?

What lake between Bolivia and Peru? What high mountain near it?

What river flows through the southern part? Into what does it flow?

By what is the northern part of Bolivia drained?

What city south-west of its capital?

What desert between the Andes and the Pacific Coast?

What volcano in the south-west? What town?

Bound the ARGENTINE REPUBLIC. What is its capital?

Name its principal cities? What large river flows through the country?

What volcanoes between the Argentine Republic and Chili?

In what part are the Pampas? What capes on the coast?

Bound PARAGUAY. What is its capital?

What two large rivers unite at the south-west corner of Paraguay?

What division of land is Paraguay? What town in the west?

Bound URUGUAY. What is its capital?

Bound CHILI. What is its capital? What cities on the coast?

In the south-west, what islands? Peninsula? Gulf?

Bound PATAGONIA. Has it any capital? Cities?

On what ocean does Patagonia border? What two bays east?

What capes on its coast? What peninsula?

What islands east of the southern part?

What islands compose the group of Terra del Fuego?

What strait between Patagonia and Terra del Fuego?

REVIEW.

CITIES AND TOWNS.

Where situated? On or near what waters?

MONTE VIDEO.	PARAMARIBO.	CARACAS.	POTOSI.
GEORGETOWN.	PERNAMBUCO.	LA PAZ.	ASPINWALL.
VALPARAISO.	POPAYAN.	BOGOTA. (bo-go-tah'.)	QUITO.
RIO JANEIRO.	MARACAIBO.	SANTIAGO. (sahn-te-ah'go.)	LIMA.
BUENOS AYRES.	PANAMA.	AREQUIPA. (ah-ra-kee'pah.)	BAHIA.
CHUQUISACA.	SANTA FE.	PARANA. (pah-rah-nah'.)	CUZCO.
CAYENNE.	COBIJA. (ko-bee'hah.)	CALLAO. (kahl-lah'o.)	PARA.

MOUNTAINS.

Where are they? In what direction do the ranges extend? Which are volcanoes?

BRAZILIAN ANDES?	ILLIMANI? 21 (eel-yah-mah'ne.)	ANDES?	ANTUCO?
COTOPAXI? 19		GERAL?	ACARAY?
CHIMBORAZO? 21	ANTISANA? 19	AREQUIPA? 20	ATACAMA?
ACONCAGUA? 23	PACARAIMA?	PICHINCHA? 16	

RIVERS.

Where do they rise? In what direction do they flow, and into what waters? Which are boundary rivers?

ST. FRANCISCO?	TOCANTINS?	PARANA?	PURUS?
TUNGURAGUA?	PUTUMAYO?	URUGUAY?	PARA?
MAGDALENA?	UCAYALI?	AMAZON?	NEGRO?
PILCOMAYO?	ORINOCO?	MADEIRA? (mah-da'rah.)	TAPAJOS?
PARAGUAY?	LA PLATA?		

CAPES.

Where are they? Into what waters do they project?

CORRIENTES?	GALLINAS?	ST. ROQUE?	BLANCO?	HORN?
ST. FRANCISCO?	ST. LORENZO?	ORANGE?	NORTH?	FRIO?

AREAS IN SQUARE MILES.	POPULATION.
Mexico, 772,850.	Mexico, 8,218,000.
Nicaragua, 58,169.	Guatemala, 1,180,000.
Honduras, 47,092.	San Salvador, 600,000.
Guatemala, 44,778.	Nicaragua, 400,000.
Costa Rica, 21,495.	Honduras, 350,000.
Balize, 14,000.	Costa Rica, 135,000.
San Salvador, 7,335.	Balize, 10,000.
Cuba, 42,383.	Cuba, 1,247,000.
Hayti, 29,000.	Hayti, 560,000.
Jamaica, 6,250.	Porto Rico, 590,000.
Porto Rico, 3,865.	

Church and Square of San Domingo, City of Mexico.

1. ***MEXICO corresponds in Latitude*** with the Great Desert of Africa, the southern half of each being in the Torrid Zone.

2. ***Its leading Characteristics*** are its high mountains and table-lands, its various climates and productions, and its precious metals.

3. ***Its Climates and Productions*** are those of the Torrid, Temperate and Frigid Zones. The hot or Torrid Region embraces the low lands along the coasts and the slopes to the height of 2500 feet; thence to the altitude of 5,000 feet, is the Temperate Region; above 8,000 feet, the cold is severe, and at the elevation of 14,800 feet, is the limit of perpetual snow (see page 39).

4. ***The Hot Region Produces*** sugar, coffee, indigo and tropical fruits. Its groves contain parrots, paroquets and other tropical birds of beautiful plumage, besides venomous snakes and insects. In summer and autumn it is exceedingly hot; and, owing to abundant rains and rank vegetation, very unhealthful.

5. ***The Temperate Region*** has an agreeable and healthful climate. Its products are grain, fruits and vegetables.

6. ***The Animals*** comprise immense numbers of cattle, wild horses, sheep, and bisons, or American buffaloes. The cattle and horses were originally introduced by the Spaniards.

7. ***The Capital*** is the city of Mexico, situated in the center of a plateau, 7,400 feet above the level of the sea (see p. 16, pars. 32, 33).

8. ***The Original Inhabitants*** were civilized Indians.

9. ***Mexico was Conquered*** by Cortez in 1520, and it remained in the possession of Spain for nearly three centuries afterwards.

10. ***The Government,*** which is a Union of several States, is in a very unsettled condition, and the resources of the country lack development.

11. ***The Population*** of Mexico is about 8 millions, more than one-half of whom are Indians; the remainder comprise whites, negroes and mixed races. The whites are of Spanish descent and are called Creoles.

12. ***CENTRAL AMERICA*** comprises five independent States or republics—Guatemala, Honduras, San Salvador, Nicaragua and Costa Rica, besides British Honduras, or Balize. The most important of these is Guatemala.

13. ***Its Surface,*** climates, productions and inhabitants are similar to those of Mexico.

14. ***The leading Exports*** are indigo, cochineal, coffee, sugar, cotton and mahogany.

15. ***THE WEST INDIES*** include the Greater Antilles, the Lesser Antilles, the Caribbee and the Bahama Islands; all, excepting some of the Bahamas, are in the Torrid Zone.

16. ***Their Surface,*** generally, is mountainous.

17. ***Their Climate*** is very hot, but tempered by the influence of the ocean. It is generally healthy from November to June; but, during the great rains, fevers are prevalent.

18. ***Hurricanes and Earthquakes*** are frequent.

19. ***The Vegetable Productions*** are rich and abundant—sugar, coffee, tobacco, cotton, corn, mahogany, dye-woods and drugs, besides pine-apples, oranges, bananas and other tropical fruits.

20. ***The Cultivation of the Soil*** is performed by negroes, who comprise the larger portion of the inhabitants.

21. ***Fish, Turtles,*** parrots, flamingoes, humming-birds, snakes, lizards and insects are numerous.

22. ***CUBA,*** the largest of the West Indies, is 750 miles long.

23. ***Its leading Manufactures*** are sugar, molasses and cigars. Its commerce is important; its chief city is Havana.

24. ***HAYTI*** comprises two republics of free colored people; the French language is spoken in the western part of the island, and Spanish, in the eastern part.

1. *SOUTH AMERICA:* its leading characteristics are its high mountains, volcanoes, plateaus, large rivers and extensive plains.

2. *It has three* mountain systems, three plateau regions, and three great river basins.

3. *The most important* mountains are the Andes, which consist of several ranges, and extend along the Pacific coast a distance of over 4500 miles.

4. *In the Andes,* are plateaus and valleys of various sizes and elevations.

5. *The Highest Peaks* are more than 20,000 feet above the level of the sea. Mt. Aconcagua, the highest, is 23,910 feet.

6. *The Largest Plateau of the Andes* is in Bolivia and Peru; it is more than 12,000 feet above the level of the ocean, and covers an area equal to that of Indiana.

The Upper Andes.—Head of the Apurimac River, one of the Sources of the Amazon.

7. *The greater part of South America* is in the Torrid Zone; consequently, Bolivia, Peru, Ecuador, and Colombia possess every variety of climate, from the tropical heat of their low plains, to the region of perpetual snow on their mountains. (See page 39.)

8. *The Countries bordering on the Pacific* are subject to earthquakes.

9. *The other Mountain Systems* of South America are north of Brazil and in the southeastern part of that country.

10. *The most extensive Plateau* of South America is in the central and southeastern portions of Brazil. Its elevation is from 2,000 to 3,000 feet above the sea level.

11. *The excessive rains of the Tropical Regions* of South America are supplied by the winds which blow from the Atlantic. As the moisture is condensed before passing the Andes, little or no rain falls between those mountains and the Pacific, which accounts for the deserts of that region. (See p. 34, pars. 41 to 45.)

12. *The Seasons* of the year are two, the wet and dry.

13. *The Lowlands* comprise, chiefly, the basins of the Amazon, Orinoco, and La Plata.

14. *The Basin or Valley of the Amazon* has an area of over 2,000,000 square miles, and includes large portions of Brazil, Ecuador, Peru and Bolivia.

15. *Its Surface* is covered with numerous streams, marshes, and dense forests, called silvas, the abode of savages, monkeys, alligators, huge serpents and venomous insects. (See page 44, paragraph 37.)

16. *The Lowlands of the Orinoco* are called llanos; those of the La Plata, pampas. In the wet season, they are covered with grass; but, in the dry season, they become desolate wastes. (See page 21, paragraphs 9 to 13.)

17. *These Lowlands* are in nearly every country in South America. For subsistence and trade, a large portion of the inhabitants depend upon their cattle and horses, which roam over the llanos and pampas, in countless herds, during the wet season.

18. *Animals* are numerous; they include the jaguar, puma, tapir, ant-eater, sloth and armadillo. Mules and lamas are used as beasts of burden.

19. *The Agricultural Products* are coffee, sugar, wheat, corn, cotton, tobacco, rice and tropical fruits. In the forests, are the chocolate and India-rubber trees, and those from which cabinet and dye-woods, Peruvian bark and other medicines are obtained.

20. *The Largest Birds* are the emu and vulture; besides these are the toucan, with parrots and humming-birds of great beauty.

21. *The Original Inhabitants* were Indians: those under the government of the Incas, whose dominions extended along the Andes from the Equator to Patagonia, had advanced to a high degree of civilization.

22. *Some of their Buildings* were of magnificent construction, especially the Temple of the Sun, which was ornamented with gold and jewels of great value.

23. *Peru* was conquered by Pizarro, about the year 1533, and remained in possession of the Spaniards for about 300 years.

24. *The present Inhabitants* of South America are Indians, whites, negroes and mixed races. More than one-half are Indians. The white inhabitants are chiefly of Spanish and Portuguese descent; Portuguese, in Brazil; British, Dutch and French, in Guiana; and Spanish, in all the other countries.

25. *The South American Governments* are republics, except Brazil, which is an empire, and Guiana, which is under the dominion of Great Britain, France and Holland.

26. *BRAZIL* is the largest country in South America; its wealth is in its fertile soil, extensive pastures, its gold, diamonds, and timber.

27. *It Produces* more than half the coffee used in the world.

28. *The Northern Part* of Brazil is subject to heavy rains and violent storms. In the south, the climate is mild and salubrious.

29. *The Largest City* in South America is Rio Janeiro, the capital and chief commercial city of Brazil.

30. *The Cities, Towns, and Cultivated Districts* of Brazil are chiefly near the coast.

31. *THE ARGENTINE REPUBLIC:* its characteristics are its forests and deserts of the north, and its vast pampas of the center and south.

32. *Its Climate* is distinguished by great summer heat, violent hail-storms and long droughts.

33. *The National Wealth* is mostly in the cattle, horses, mules and sheep, which are reared, in millions, on the pampas; the leading exports are wool, hides and skins.

34. *Patagonia* is mostly a barren region, claimed by Chili and the Argentine Republic.

BRITISH ISLES

EXERCISES ON THE MAP.

Name the countries of the BRITISH ISLES. Which is the **largest?** Bound England, Scotland, Ireland, Wales.

In what island are England, Scotland, and Wales comprised? *Ans.—Great Britain.*

What rivers of England flow easterly? Westerly?

What is the capital of England? What city at the mouth of the Mersey? What city east of Liverpool? What three north-east of Manchester? What towns in the south?

What hills in Scotland? What firths on the eastern coast? Western? What heads or capes on the coast? What group of islands north? North-west? What is the capital of Scotland? What city west of Edinburgh? What is the capital of Ireland? What cities in the north? East? South? West?

What rivers flow northerly? Easterly? Southerly, and south-westerly? What lochs in the north of Ireland? Near the centre? What bays west and south-west? What island in the Irish Sea? What island north-west of Wales?

What is the length of Great Britain? Of California?

What is the greatest breadth of each?

REVIEW.

CITIES AND TOWNS.

In what part of what county are they? On or near what water?

LONDON,	DUBLIN,	ABERDEEN,	SOUTHAMPTON,
MANCHESTER,	LEEDS,	PORTSMOUTH,	SWANSEA,
LIVERPOOL,	BRISTOL,	NORWICH,	GALWAY,
GLASGOW,	BELFAST	LIMERICK,	WATERFORD.
BIRMINGHAM,	HULL,	PLYMOUTH,	PERTH,
EDINBURGH, (ed'in-bur-ruh)	CORK,	PAISLEY,	LONDONDERRY.

ISLANDS.

Where are they? By what waters are they surrounded?

ORKNEY IS.?	ANGLESEA? (ang'gle-se-a)	WIGHT?	LEWIS?	MAN?
HEBRIDES?		ARRAN?	SCILLY IS.?	SKYE?

RIVERS.

Where do they rise? In what directions do they flow, and into what waters?

THAMES?	SEVERN?	BANN?	BOYNE?	TAY?
SHANNON?	MERSEY?	HUMBER?	TRENT?	AVON?
LIFFEY?	OUSE? (ooz)	TWEED?	FOYLE?	DEE?
BLACKWATER?		CLYDE?		

BAYS.

Where are they? Into what waters do they open?

THE WASH?	PENTLAND FIRTH?	FIRTH OF TAY?
FIRTH OF FORTH?	LOCH FOYLE?	GALWAY?
DONEGAL?	LOCH LINNHE?	MURRAY FIRTH?

Areas Compared.

Sq. miles.
England and Wales, 58,320.
Georgia (U. S.), 58,000.

Scotland (inc. Islands), 31,324.
Maine, 31,766.

Ireland, 32,509.
Indiana, 33,809.

The Eastern Part of Edinburgh Looking South.—The Palace of Holyrood.—Salisbury Crags and Arthur's Seat in the Distance.

Populations Compared.

England and Wales, 22,704,100.
Farther India, 30,000,000.

Scotland (inc. Islands), 3,358,613.
New England States, 3,487,924.

Ireland, 5,402,759.
Pennsylvania and Ind., 5,202,687.

1. ***THE BRITISH ISLES*** comprise Great Britain and Ireland, with many small islands near their coasts.

2. ***They are Situated*** in the centre of the land hemisphere, and between the same parallels of latitude as Labrador and the southern half of Siberia. (See Land Hemisphere, p. 50.)

3. ***Their Climate*** is mild and moist, owing to the influence of the west winds, which blow over the Gulf Stream. See pages 35 and 36.)

4. ***Their Area*** is less than that of New Mexico; and no part of their surface is more than 100 miles from the coast.

5. ***The Largest European Island*** is Great Britain, the most important in the world. It comprises England, Scotland and Wales. The second in size, is Ireland.

6. ***Their nearest approach to Continental Europe*** is at the Strait of Dover, (21 miles wide) between England and France.

7. ***The most Northern*** of the British Isles are the Shetland Islands.

8. ***The British Empire comprises*** the British Isles and extensive possessions in America, Asia, Africa, and Oceanica.

9. ***ENGLAND*** possesses fine pastures, well cultivated farms, beautiful landscapes, and mines which yield immense wealth.

10. ***Its Mountains*** are in the west and northwest, the highest being about 3,000 feet high. Its lowest lands are in the east.

11. ***In the Northwest*** are the mountains, lakes, and waterfalls of England, which are celebrated for the beauty of their scenery. The highest mountains are Sca Fell, Helvellyn and Skiddaw, each over 3,000 feet. The largest lake—Windermere—is about 10 miles long by 1 mile wide.

12. ***The Mineral Products*** include coal, iron, copper, lead, tin, and salt. The coal fields are, chiefly, in the northern counties.

13. ***The Agricultural Products*** are chiefly wheat, rye, barley and oats. Indian corn and the grape do not thrive.

14. ***England excels*** every other country in the world in the extent of its commerce and manufactures; this is largely due to its facilities for navigation, and the abundance of its coal and iron.

15. ***Its Cotton, Woolen and Iron*** manufactures are immense.

16. ***England*** is the most densely populated country in Europe, except Belgium.

17. ***London***, the capital, excels every other city in the world, in wealth, population and commerce.

18. ***Liverpool***, next in size to London, is a celebrated commercial city; next in rank, are Manchester, noted for its cotton manufactures, and Birmingham, for hardware.

19. ***WALES*** is a mountainous country, and well adapted to grazing; its manufactures are woolen goods.

20. ***SCOTLAND:*** its distinguishing features are its broken coast line, its rugged surface, and its numerous lakes.

21. ***The Highlands***, which cover the northern half of Scotland, comprise the Grampian Hills and other ranges.

22. ***They are Remarkable*** for the wild scenery of their naked rocks and precipices, narrow glens, lakes and waterfalls, and their desolate moorlands, which are covered with heath and bog.

23. ***The Highest of the Grampian Hills*** is Ben Nevis, 4406 feet (the highest mountain in Great Britain).

24. ***The Largest Lake*** in Great Britain is Loch Lomond.

25. ***The Lowlands***, which are comparatively level, contain good agricultural lands, and the bulk of the population. Here the people are chiefly engaged in agriculture, manufactures and commerce. The Highlanders are chiefly shepherds.

26. ***The Agricultural Products*** are oats, wheat, barley, potatoes and turnips. The minerals are coal, iron and lead.

27. ***The Capital*** is Edinburgh, and the chief city, Glasgow.

28. ***IRELAND*** is hilly or mountainous along the coast, and generally level in the interior. The highest mountains are a little over 3000 feet high.

29. ***Ireland is chiefly*** a grazing and agricultural country; oats, wheat, potatoes, barley and flax are successfully cultivated.

30. ***The Principal Manufacture*** is linen.

31. ***Dublin*** is the capital and chief city. The other large cities are Belfast, Cork, and Limerick.

32. ***Celebrated for Beautiful Scenery***, are the Lakes of Killarney, and the region surrounding them.

33. ***The Government*** of the British Isles is a limited monarchy. The legislative body, or parliament, is composed of the House of Lords and the House of Commons: it comprises members from England, Scotland, Ireland and Wales.

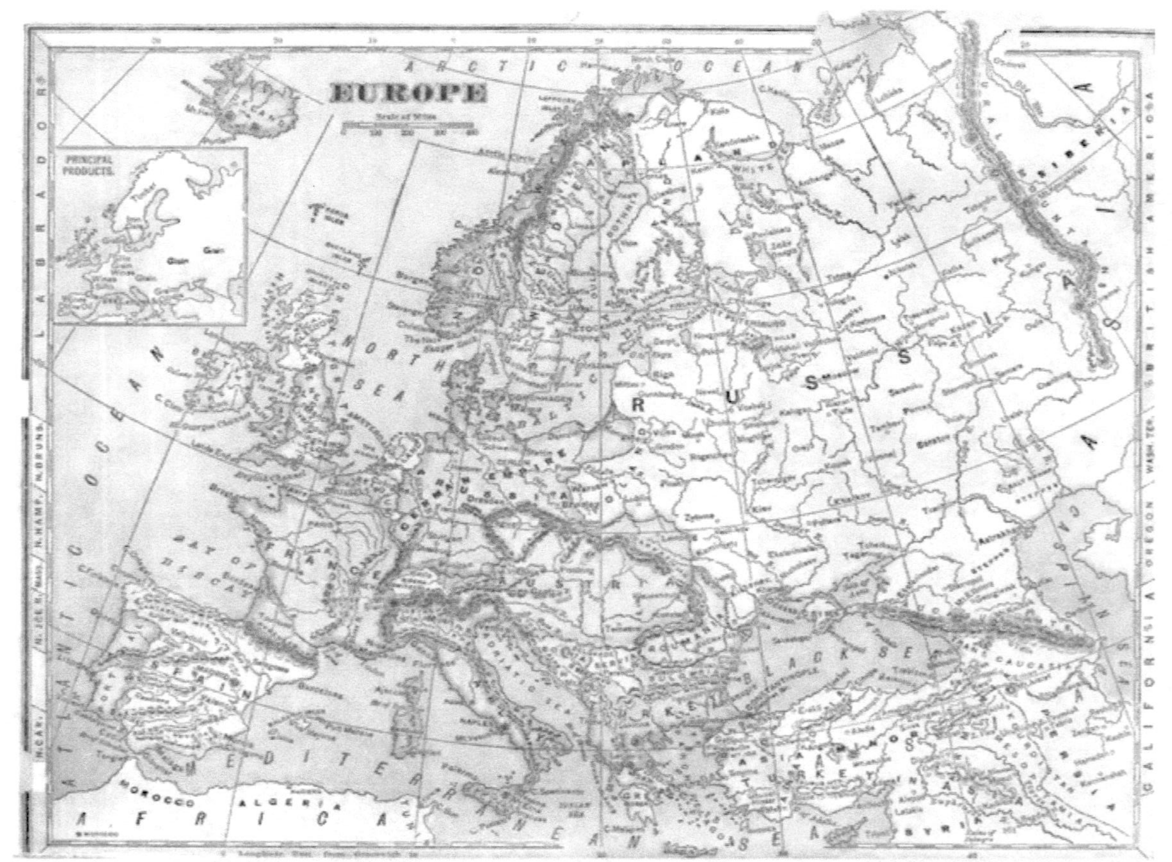
EUROPE
PRINCIPAL PRODUCTS.
ARCTIC OCEAN
ATLANTIC OCEAN
NORTH SEA
BAY OF BISCAY
MEDITERRANEAN SEA
BLACK SEA
CASPIAN SEA
ADRIATIC SEA
RUSSIA
FRANCE
SPAIN
AUSTRIA
TURKEY
ASIA MINOR
TURKEY IN ASIA
PERSIA
SYRIA
CAUCASIA
SIBERIA
URAL MOUNTAINS
ASIA
AFRICA
MOROCCO
ALGERIA
LABRADOR
BRITISH AMERICA
OREGON
CALIFORNIA

SECTION OF EUROPE FROM THE BAY OF BISCAY TO THE CASPIAN SEA.

EXERCISES ON THE MAP.

What parallel of latitude passes through the northern part of Europe?

In what zones is Europe? In what zone is the greater part? What ocean north of Europe? West? What sea south? What inlets in Europe from the Arctic Ocean? From the Atlantic? From the Mediterranean Sea?

What is the largest country in Europe? What sea coast has Russia? Great Britain? France? Spain? Italy? Turkey? Austria? Prussia? Holland?

What country in Europe has no sea coast?

What countries comprise peninsulas?

What two rivers flow into the Caspian Sea? Into the Black Sea?

What river flows into the Sea of Azov? Into what does the Sea of Azov open? What connects the Black Sea with the Mediterranean?

What mountains extend from the Black to the Caspian Sea? What mountains and river between Europe and Asia? What mountains between Norway and Sweden? In the eastern part of Austria? In Turkey?

In what mountains does the Danube rise?

What large tributaries has the Danube? What rivers rise in the Alps? Which flows north? South? East?

What three seas receive water from the Alps?

In what part of Russia is Finland? By what waters is it embraced?

What three gulfs open into the Baltic Sea? What gulf into the White Sea?

Where is Lapland? In what zone is it?

What lake empties into the Gulf of Finland? Into Lake Ladoga? (lad'o-gah.)

What city on the outlet of Lake Ladoga? At the mouth of the Dwina?

What city at the mouth of the Volga? On the Gulf of Riga? On the north-western part of the Black Sea? On the Caspian Sea?

What city near the center of Russia? In Poland?

What cities of Europe on or near the parallel of 60 degrees north latitude? Of 50 degrees? Of 40 degrees?

What cities in the United States on or near the parallel of 40 degrees?

What lands and water of Europe between the parallels of 50 and 60 degrees?

What land and water of North America between those parallels?

What is the most southern country in Europe? Its southern cape?

What is the capital of France? Great Britain? Scotland? Ireland? Italy? Spain? Portugal? Germany? Prussia? Belgium? Holland? Denmark? Austria? Turkey? Switzerland? Greece? Russia?

What large island south-east of Greece? East of Candia? What island south of Italy? What two west?

What group east of Spain? West of Norway? North-west of Scotland?

What three groups north of Scotland? Bound Denmark, Spain, and Portugal? What rivers in Spain?

Bound **RUSSIA**. What rivers in Russia flow north? South? South-east?

Bound **NORWAY**. Are there any long rivers in Norway?

Bound **SWEDEN**. In what direction do the rivers of Sweden flow? In what mountains do they rise? Into what waters do they flow?

ROUTES OF TRAVEL.

In what directions and on what waters would you sail from Liverpool to Marseilles? From Marseilles to Asia Minor? Copenhagen to St. Petersburg? Paris to Rome? Odessa to Athens? Naples to Lisbon? Havre to Marseilles? Edinburgh to London? London to Dublin? Copenhagen to Iceland? Christiania to Archangel? Constantinople to the Sea of Azov?

What is the distance across Europe from Ireland to the Ural Mountains? Across the United States from east to west?

What is the length of the Mediterranean Sea? Of the Gulf of Mexico? Of the Black Sea? Of Lake Superior?

What is the distance in a straight line between London and Liverpool? Between London and Dublin? Between London and Paris? Between Paris and Marseilles?

State the length and breadth of Greece? Of Ohio?

Which of the United States is larger than any European country except Russia?

REVIEW.

MOUNTAINS.

Where are they? In what directions do the ranges extend?

Mt. Elboorz? 17	Cantabrian? 4	Scandinavian?
Mt. Blanc? 10	Apennines? 5	Pyrenees?
Alps? 10	Mt. Vesuvius?	Ural?
Carpathian? 3	Mt. Hecla?	Mt. Etna?
Caucasus? 16	Sierra Nevada? 8	Balkan?

RIVERS.

Where do they rise? What courses do they take? Into what do they flow?

Volga? 20	Petch'ora?	Douro?	Ebro?
Danube? 17	Rhine?	Drave?	Save?
Ural? 11	Rhone?	Guadiana?	Seine?
Vistula?	Dwina?	Tagus?	
Dnieper? 12 (ne'per.)	Garonne? (gah-ronn'.)	Guadalquiver? (gwah-dal-quiv'er.)	Loire? (lwahr'.)

SEAS, GULFS, AND BAYS.

Where are they? Into what waters do they open?

Mediterranean?	Baltic S.?	G. of Onega?
Ad'riatic S.?	Black S.?	North S.?
B. of Biscay?	G. of Finland?	S. of Az'ov.?
G. of Bothnia?	White S.?	G. of Taranto?
Caspian S.?	G. of Lyons?	
Archipelago? (ar-ke-pel'ah-go.)	S. of Marmora? (mar'mo-rah.)	G. of Riga? (re'gah.)

STRAITS AND CHANNELS.

What lands do they separate? What waters do they connect?

Gibraltar?	Otranto?	St. George's C.?
English C.?	Yenikale?	Dardanelles?
Dover?		

ISLANDS.

Where are they? By what waters are they surrounded?

British Is.?	Shetland Is.?	Cyprus?
Loffoden Is.?	Sardinia?	Corsica?
Ionian Is.?	Balearic Is.?	Candia?
Faroe Is.?		Sicily?

CAPES.

Where are they? Into what waters do they project?

Spartivento?	Passaro?	North?
Matapan?	Finisterre?	Clear?
St. Vincent?		Ortegal?

CAPITAL CITIES.

Of what countries are they capitals? Where situated?

LONDON,	MADRID,	COPENHAGEN,
PARIS,	DUBLIN,	STOCKHOLM,
CONSTANTINOPLE,	LISBON,	FRANKFORT,
BERLIN,	EDINBURGH,	THE HAGUE,
ST. PETERSBURG,	BRUSSELS,	ATHENS,
VIENNA,		BERN,

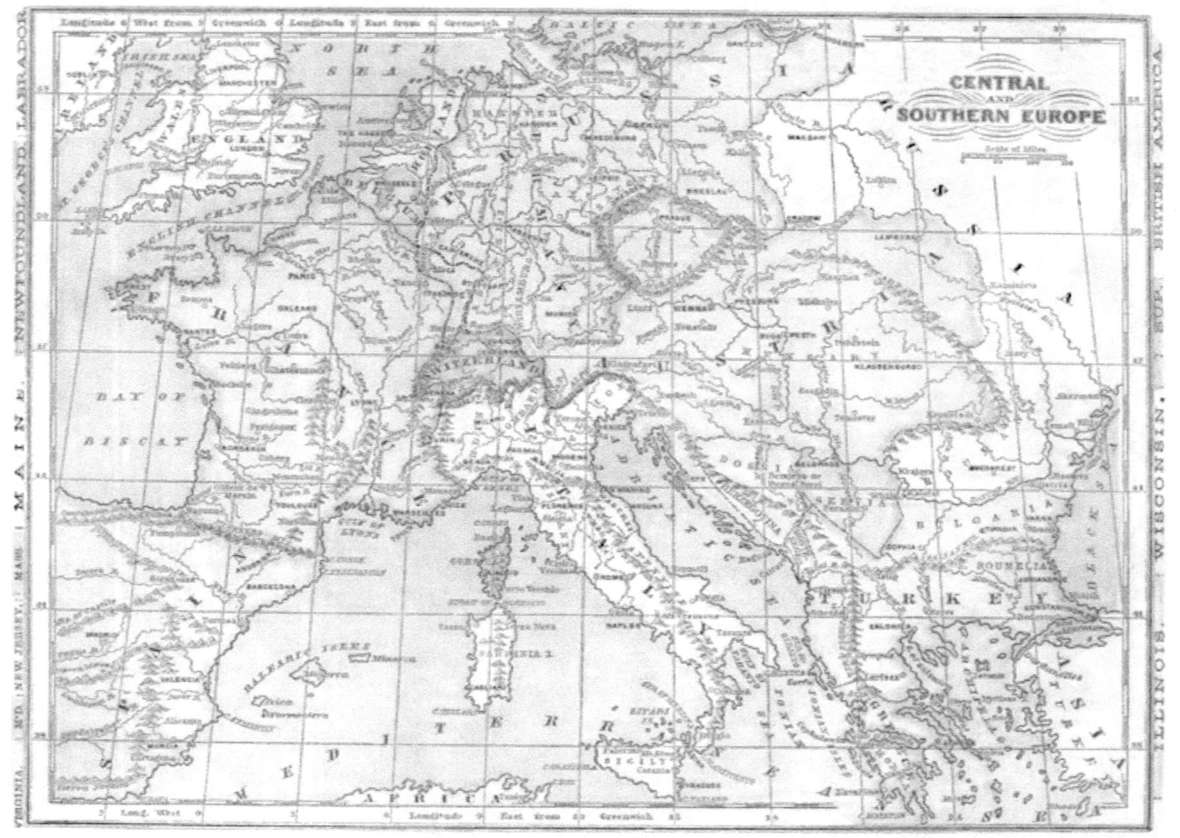
CENTRAL AND SOUTHERN EUROPE
Scale of Miles
NORTH SEA
BALTIC SEA
ENGLAND
IRISH SEA
ENGLISH CHANNEL
BAY OF BISCAY
FRANCE
SPAIN
BALEARIC ISLES
SWITZERLAND
ITALY
SICILY
SARDINIA I.
ADRIATIC SEA
IONIAN SEA
IONIAN ISLES
AUSTRIA
HUNGARY
RUSSIA
TURKEY
BULGARIA
ROUMELIA
SERVIA
BOSNIA
GREECE
ARCHIPELAGO
ASIA
BLACK SEA
AFRICA
MEDITERRANEAN
NEWFOUNDLAND, LABRADOR.
MAINE.
MASS.
NEW JERSEY.
M'D.
VIRGINIA.
BRITISH AMERICA.
SUP.
WISCONSIN.
ILLINOIS.

EXERCISES ON THE MAP.

By what is **FRANCE** bounded on the north? East? South? West?

What mountain ranges in France? What high mountain in the eastern part?

In what direction does the surface in the northern part slope?

Name the rivers on the northern slope.

What river drains that part of France east of the Cevennes (*se-ven'*) Mountains?

What drains the valley between the Cevennes and Auvergne (*o-vairn'*) Mountains?

What rivers drain the western part of France?

What river rises in France and flows through Belgium and Holland?

What tributary of the Rhine has its source in France?

Name the principal tributaries of the Rhone. Seine. Loire.

In what direction does the north-west coast of France extend? Its north-east boundary?

What is the capital of France? What city at the mouth of the Seine? What two cities on the Loire? On the Garonne?

What city is farthest west? What three on the coast of the Mediterranean?

What small republic among the Pyrenees? *Ans.* Andorra.

Bound **BELGIUM.** What is its capital?

By what two rivers is Belgium drained?

Bound **HOLLAND.** What is its capital?

What inlet from the North Sea extends into the land?

What large river flows through Holland?

By what is **PRUSSIA** bounded on the north? East? South? West?

In what part of Prussia is Hanover? Holstein?

What mountains on its southern border?

In what general direction does the land of Prussia slope?

What rivers flow into the North Sea? Baltic Sea? Gulf of Dantzic?

What is the capital of Prussia? Name the chief cities of Prussia.

Bound **AUSTRIA.** What is its capital?

What rivers rise on the eastern slope of the Carpathian Mountains?

What is the largest river in this part of Europe?

Near what large river is the source of the Danube?

What tributaries of the Danube have their sources in the Alps? In the Carpathian Mountains?

Name the chief cities of Austria?

As a result of the war with Prussia in 1866, Austria was compelled to withdraw from the Germanic Confederation.

The **GERMAN EMPIRE** comprises the Kingdoms of Prussia, Bavaria, Wurtemberg, and Saxony, and the smaller German States. In the war of 1870-71, between France and Germany, the French were defeated and compelled to cede that portion of their territory which lies west of Baden, to Germany.

What is the capital of the Kingdom of Wirtemberg? Of Saxony? *Ans.—Dresden.* Of the Duchy of Baden (*bah'den*)?

What country in Europe has no sea coast?

Bound **SWITZERLAND.** What is its capital?

Mention its principal cities.

What four large rivers have their sources in the mountains around Switzerland and flow into four different seas?

What can you say of the large rivers of Central Europe? *Ans.* They all rise in or near Switzerland and flow in different directions; their sources are near together, but their mouths widely separated from each other.

Bound **ITALY.** What is its capital?

What mountains on the northern boundary? What chain extends through Italy? What two rivers in the north?

What three divisions in the north? Name their chief cities. What large division in the south? What is its principal city? What two large islands are included in the Kingdom of Italy? *Ans.* Sicily and Sardinia.

What small republic north-east of Florence? *Ans.* San Marino. What important port south-east of San Marino?

What gulf north-west of Italy? Sea and gulf south-east?

What strait between Italy and Sicily? Corsica and Sardinia?

What volcano near the city of Naples? In Sicily?

What two cities in Sicily? In Corsica and Sardinia?

What is the southern cape of Italy? The south-eastern?

What is the southern cape of Sicily? Of Sardinia?

What is the northern cape of Corsica?

What islands south of Sicily? East of Spain?

Bound **TURKEY.** What is its capital?

What mountain range in the south-west? North-east?

What great river on the north flows into the Black Sea?

What three independent States were, until 1878, subject to Turkey? *Ans.* Roumania, Servia, and Montenegro.

What provinces formerly subject to Turkey are now occupied by Austria? *Ans.* Bosnia and Herzegovina.

What portion of Turkey was ceded to Russia? *Ans.* That part lying east and north of the Pruth River.

What Turkish island in the Mediterranean was ceded to England? *Ans.* Cyprus.

What city in Servia at the mouth of the Save River? What city in Roumania north of the Danube? Name the chief cities in Turkey?

Bound **GREECE.** What is its capital?

What gulf extends far into Greece? What peninsula is formed by it? What island east of Greece? What group west?

REVIEW.—MOUNTAINS.

Where are they? In what directions do the ranges extend?

Moravian?	Castile?	Sierra Morena?
Bohemian?	Pindus?	Riesen Gebirge?
Cevennes?	Auvergne?	Erze Gebirge?

RIVERS.

Where do they rise? In what directions do they flow, and into what waters?

Meuse?	Weser?	Elbe?	Dniester?
Adige?	Rhine?	Pruth? —	Moselle?
Danube?	Oder?	Po?	Wartha?
Spree?	Saone?	Iser?	Dordogne?

SEAS, GULFS, AND BAYS.

Where are they? Into what waters do they open?

Ionian Sea?	G. of Gen'oa?	G. of Dantzic?
G. of Lepanto?	G. of Salonica?	Zuyder Zee?

STRAITS.

Between what lands are they?

Messina? Bonifacio?

ISLANDS.

Where are they? By what waters are they surrounded?

Sardinia?	Sicily?	Corsica?
Negropont?	Minorca?	Ivica?
Rhodes?	Zante?	Corfu?
Ionian Is.?	Formentera?	Majorca?
Scilly Is.?	Jersey?	Cephalonia?
Balearic Is.?	Lipari Is.?	Guernsey?

CAPES.

Where are they? Into what waters do they project?

Land's End?	St. Martin?	Corso?
La Hogue?	Passaro?	St. Sebastian?
Matapan?	Teulada?	Di Leuca?
Spartivento?		

CITIES AND TOWNS.

Where are they? On or near what waters?

LUCERNE,	MODENA,	WARSAW,	BREST,
BARCELONA	GENEVA,	HANOVER,	ORLEANS,
VENICE,	LYONS,	DANTZIC,	SALONICA,
HAMBURG,	SYRACUSE,	PESTH,	TOULON,
STUTTGART,	BRESLAU,	PARMA,	MILAN,
VALENCIA,	NAPLES,	ANCONA,	BOLOGNA,
ANDORRA,	BRUSSELLS,	SAN MARINO	PRAGUE,
MARSEILLES,	BORDEAUX,	MUNICH,	TURIN,
			BUDA,
ADRIANOPLE,	TOULOUSE,	AJACCIO,	LEIPSIC,
CAGLIARI,	MESSINA,	HAVRE,	NANTES,
BUCHAREST,	CRACOW,	NICE,	CARLSRUHE,

The Palace of Versailles, near Paris; Now Used as a Historical Museum

14. ***FRANCE*** corresponds in latitude with the New England States, the Great Lakes, Minnesota and Oregon.

15. ***Its Mountains*** are on its eastern and southern borders; they are noted for the grandeur of their scenery.

16. ***The River Valleys*** present beautiful landscapes.

17. ***Its Climate,*** soil, and agricultural productions are varied; the cold of winter increases according to the distance inland. The southeastern section is noted as a winter resort for invalids.

18. ***The Most Important Agricultural Products*** are grain and grapes, the former flourishing in the north, the latter in the south.

19. ***The Commerce and Manufactures*** of France are very important; the latter comprise silk, woolen, linen and cotton goods.

20. ***The Largest City*** in France is Paris, the capital and third city in size in the world; London and Pekin being the largest.

1. ***EUROPE*** is remarkable for its great length of coast line, its great plain, mountain and river scenery, and powerful nations.

2. ***Its Northern Half*** is mostly level; its southern half, mountainous.

3. ***Its Great Plain*** comprises two-thirds of its area, covering Russia, Prussia, Denmark and Holland.

4. ***Its Most Celebrated Mountains*** are the Alps; Mount Blanc, the highest mountain in Europe, is 15,810 feet above the level of the sea.

5. ***The Rivers*** which are celebrated for the beauty of their scenery are the Rhine, Rhone, Seine, Loire and Danube. The Rhine is noted for its vine-covered hills, picturesque villages, its crags and ancient castles. Its most interesting portion is in Prussia, between the towns of Bonn and Mayence.

6. ***The Foreign and Inland Commerce*** of Europe is very extensive, owing largely to the numerous indentations of its coast, and its network of rivers, canals, and railroads.

7. ***The Climate*** of Western and Southern Europe is greatly modified by the winds which blow over the Gulf Stream, and the Mediterranean Sea. It corresponds to that of the Pacific coast of the United States. (See p. 37, paragraphs 21, 31 and 38.)

8. ***The Warm Zone*** of Europe embraces Spain, Italy, Turkey, Greece, and Southern France. The winters are short, frost and snow, rare, with but slight interruption to vegetation. Here flourish the vine, orange, lemon, citron, fig and olive. The heat of this region is often greatly increased by the hot winds from Africa.

9. ***The Central, or Temperate Region,*** except along the western coast, has long, cold winters, with considerable snow. Here grain is largely produced.

10. ***The Northern and Eastern Parts*** of Europe are remarkable for their severe winters, and the rapid growth of vegetation in their warm, short summers.

11. ***Edinburgh and Moscow,*** although in the same latitude, have very different climates; one is oceanic and mild, the other continental and excessive.

12. ***The Governments*** of Europe are mostly hereditary monarchies.

13. ***The Leading European Nations*** are Great Britain, Prussia, Russia, and France.

21. ***PRUSSIA*** has a northern slope. Its surface is even, except in the south and west.

22. ***THE GERMAN EMPIRE*** was formed by the union of all the Kingdoms, States, and Free Cities of North and South Germany.

23. ***The King of Prussia*** became Emperor of Germany.

24. ***The Largest City*** in Germany is Berlin, the capital. Its population is larger than that of Philadelphia.

25. ***Other Large Cities*** are Hamburg, Breslau, Munich, Dresden, and Cologne. Strasburg and Metz were lately ceded by France to Germany.

26. ***The Leading Products*** of Germany are grain, flax, tobacco and wine, linen, cotton, woolen and silk goods, coal and the useful metals.

27. ***DENMARK and HOLLAND*** have a low, flat surface and a moist climate. Their principal products are grain, cattle, butter, etc.

28. ***They Correspond in Latitude*** with Labrador and Hudson's Bay.

29. ***The Surface of a Large Portion of Holland*** is below the level of the sea and several of its rivers; the water from them being prevented from overflowing the land, by embankments. The country is drained by means of its numerous canals, into which the water is raised by windmills and steam engines.

30. ***The Foreign Possessions*** of France and Holland in Asia, Africa and America, are important.

31. ***The Most Important Cities*** of Denmark and Holland are Copenhagen and Amsterdam, their capitals.

32. ***BELGIUM*** is the most thickly settled, and the best cultivated country in Europe.

33. ***It is Rich*** in agricultural, mineral and manufacturing products. Coal and iron are abundant. Its manufactures are Brussels carpets, linens, laces, cotton, woolen, and silk goods.

34. ***Its Capital*** and chief city is Brussels.

35. ***France, Holland and Belgium*** together cover an area about equal to that of Texas.

St. Peter's, the most celebrated cathedral in the world. Its length is 697 feet; its width, 445 feet; and its height, 448 feet. On the right of the picture appears the Palace of the Vatican, which is the residence of the Pope. It contains more than 4000 rooms. Its galleries are remarkable for their magnificence, their treasures of art, a museum, and a library.

1. *AUSTRIA* is remarkable for its mountain scenery, its great central plain of Hungary, its extensive forests, its mineral wealth, and the variety of the races which compose its inhabitants.

2. *Its Productions* are grain, grapes, hops and hemp, with olives, oranges, lemons and rice.

3. *The Largest Cities* in Austria are Vienna—its capital, Prague and Pesth.

4. *RUSSIA* is remarkable for its great area, severe climates, vast forests and steppes (grassy plains).

5. *It is Chiefly* an agricultural and grazing country; its yield of wheat and other grains is abundant.

6. *Its Largest Cities* are St. Petersburg—the capital, Moscow and Warsaw.

7. *SWEDEN AND NORWAY*, now under one sovereign, form the Scandinavian Peninsula; each has its own laws and legislature.

8. *The Surface* of Norway is exceedingly mountainous, while that of Sweden is comparatively level.

9. *The Coasts* of both are lined with islands. The Norwegian coast is remarkable for its numerous inlets, called fiords.

10. *The Climate* of Sweden is remarkable for its long, cold winters, and its short, hot summers; while that of Norway is milder, owing to the influence of the Gulf Stream.

11. *The Peninsula is remarkable* for its forests, pastures, and its stores of iron and copper. Ship-building, commerce, the lumber trade, and the fisheries, are important.

12. *Stockholm*, their capital and chief city, is built partly on several islands.

13. *SWITZERLAND*, an inland republic, is the most mountainous country in Europe.

14. *It is remarkable* for its grand and picturesque scenery, especially that of its mountains, glaciers, valleys, lakes and waterfalls.

15. *Its Area and Population* are about double those of Massachusetts. More than half the inhabitants are Germans, who occupy the northeastern cantons or states; in the west are French, and in the south, Italians.

16. *Its Capital* is Berne; its chief city, Geneva.

17. *ITALY* is celebrated for its mild winters and clear sky, its mountain and lake scenery, its volcanoes and ancient ruins.

18. *Rome*, the ancient capital of the Roman Empire, is, next to Jerusalem, the most celebrated city in the world. It was once "the mistress of the world," but now is noted chiefly for its ruins, the most famous of which are the Coliseum and the Roman Forum. Its celebrated Cathedral, St. Peters, is the largest church in the world. Rome was founded more than 700 years B. C., and was in the height of its power at the beginning of the Christian era.

19. *The Largest City* is Naples, near which are Vesuvius and the ruins of Pompeii (pom-pay'e) and Herculaneum.

20. *The Principal Volcanoes*, besides Vesuvius, are Mt. Etna, in Sicily, and Strom'boli, of the Lip'ari group, all Italian islands. Corsica, the birthplace of Napoleon I., belongs to France.

21. *SPAIN AND PORTUGAL* correspond, in latitude, with Missouri and Iowa; and in area, with Missouri, Iowa and Minnesota.

22. *Their Surface* consists of plateaus and mountains. Their climate is mild on the coasts.

23. *The Highlands* are dry and unproductive; but the vine, tobacco, sugar cane, and tropical fruits flourish in the lower grounds. Merino sheep in vast numbers are reared for their wool, which is the finest in Europe.

24. *Their Largest Cities* are the capitals, Madrid and Lisbon.

25. *THE TURKISH or OTTOMAN EMPIRE* lies in Europe, Asia, and Africa.

26. *The Surface* of Turkey in Europe is mountainous and undulating; the climate, variable. Grapes and grain flourish.

27. *Many of the* people are Mohammedans; the emperor is called the Sultan, and his court, the Sublime Porte.

28. *Constantinople*, the capital, is, next to London and Paris, the largest city in Europe.

29. *GREECE* is noted for its fine scenery and climate, and its delicious fruits, especially currants, olives, and grapes.

30. *Athens*, its capital, and long one of the most celebrated cities in the world for literature and science, was founded more than 1300 years B. C.

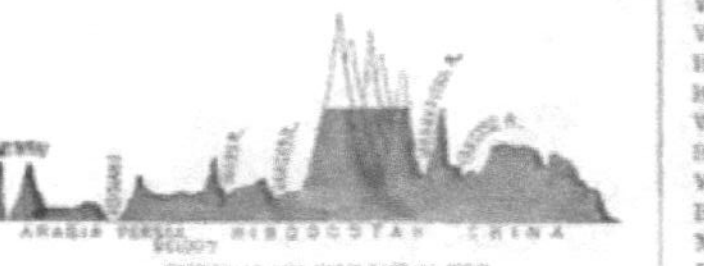

EXERCISES ON THE MAP.

What part of **ASIA** is most elevated? What rivers have their sources in the elevated regions?

What circle crosses the northern part of Asia? What tropic the southern part?

In what three zones is Asia? What countries partly within the Torrid Zone?

What countries partly within the Frigid Zone? Which are entirely within the Temperate?

To what empire does Siberia belong? What empire in the eastern part of Asia?

Name the mountain chains of Asia. The rivers of Siberia.

What mountains form the watershed of these rivers?

What rivers of Asia have an easterly course? Southerly?

In what part of Asia do the rivers have no connection with the ocean?

Into what do these rivers flow? Are the waters of the seas and lakes of this region fresh, or salt? Why?

What seas border on the west of Asia? On the east? What sea and bay south?

What two gulfs open into the China Sea? What gulfs and sea open into the Arabian Sea?

What connects the Red Sea with the Arabian? The Persian Gulf with the Arabian?

What peninsulas in the southern part of Asia? Eastern?

In what direction do they project?

Name the deserts of Asia. In what countries are they?

Bound **SIBERIA**. Name its principal cities. What city in Europe is its capital?

Bound **CHINESE EMPIRE**. Name its divisions. Its capital.

What cities in China? What great structure on its northern boundary?

Bound **FARTHER INDIA**. What are its divisions? Name the capital of each division.

What cape on the south-east? West? Island east? South?

Bound **HINDOOSTAN**. What is its capital?

In what mountains do its large rivers rise? Which have deltas?

What mountains are parallel with its eastern coast? Western?

In what zone is the northern half of Hindoostan? Its southern half?

What cities of Hindoostan north of the Tropic of Cancer? South?

What large island south-east? Groups south-west?

What is the southern cape of Hindoostan?

Bound **ARABIA**. Name its chief cities. Its deserts.

Bound **TURKEY**. What city in Europe is its capital?

What two rivers in Turkey? What mountains? Cities?

Bound **PERSIA**. Mention its capital. Its cities.

What mountains in Persia? What desert?

Bound **TURKESTAN**, **AFGHANISTAN**, **BELOOCHISTAN**. Mention their capitals.

By what mountains are these two countries separated?

What empire composed of islands east of Chinese Empire? *Ans. The Empire of Japan or Nippon.*

Which is the largest of the Japan Islands?

Which of the Japan Islands is north of Hondo? What two south?

What is the capital of Japan? Name the principal cities.

What island north of Yezo? What chain of islands between Yezo and Kamtschatka?

What island south-east of China? South?

What group south-east?

What islands and sea north-west of Asia?

What is the extent, in miles, of Hindoostan from north to south? Of the United States? Of the Chinese Empire? Of the Chinese Empire from east to west? Of the United States from east to west?

REVIEW.

MOUNTAINS.

Where are they? In what direction do the ranges extend?

HIMALAYA? 29*	ARARAT?	CAUCASUS?
ELBURZ?	STANOVOI?	ALTAI?
HINDOO COOSH?	EASTERN GHAUTS?	TAURUS?
KUEN LUN?	WESTERN GHAUTS?	URAL?

PENINSULAS.

Where are they? By what waters are they enclosed? In what direction do they project?

HINDOOSTAN?	FARTHER INDIA?	COREA?
KAMTSCHATKA?	MALAY?	

ISLANDS.

Where are they? By what waters are they surrounded?

SAGHALIEN?	LADRONE IS.?	FORMOSA?
NOVA ZEMBLA?	SUMATRA?	SHIKOKU?
KURILE IS.?	HAINAN?	JAPAN IS.?
LACCADIVE IS.?	NICOBAR IS.?	YEZO?
MALDIVE IS.?	HONDO?	LUZON?
PHILIPPINE IS.?	KIUSHIU?	CEYLON?

SEAS, GULFS, AND BAYS.

Where are they? Into what waters do they open?

ARABIAN S.?	YELLOW S.?	B. OF BENGAL?
CASPIAN S.?	BLACK S.?	CHINA S.?
PERSIAN G.?	RED S.?	G. OF SIAM?
G. OF CAMBAY?	KARA S.?	JAPAN S.?
S. OF KAMTSCHATKA?	ARAL S.?	G. OF CUTCH?
	EASTERN S.?	S. OF YEZO?
OKHOTSK S.?	CELEBES S.?	G. OF TONQUIN?

CAPES.

Where are they? Into what waters do they project?

NORTHEAST?	EAST?	NEGRAIS?
COMORIN?	CAMBODIA?	GUARDAFUI?
LOPATKA?	PRINCE OF WALES?	

RIVERS.

Where do they rise? What courses do they take? Into what do they flow?

LENA?	NERBUDDAH?	TIGRIS?
HOANG HO?	CASHGAR?	INDUS?
CAMBODIA?	MEINAM?	AMOO?
IRTISH?	IRRAWADDY?	ANADIR?
BRAHMAPUTRA?	YENISEI?	AMOOR?
YANGTSE KIANG?	OBI?	EUPHRATES?

CITIES AND TOWNS.

Where situated? On or near what waters?

TOKIO (Yedo),	ISPAHAN,	OKHOTSK,	LASSA,
PEKIN,	MANDELAY,	SMYRNA,	LAHORE,
CANTON,	MALACCA,	BANKOK,	OMSK,
CALCUTTA,	LUCKNOW,	SHIMODA,	TEFLIS,
NANKIN,	CANDAHAR,	BOMBAY,	MUSCAT,
MADRAS,	HYDERABAD,	KELAT,	MOCHA,
JERUSALEM,	DAMASCUS,	MECCA,	HUE,
FOUTCHEOU,	YAKOUTSK,	SHANGHAI,	TOMSK,
KIOTO,	NICOLAIEV,	BOKHARA,	CABUL,
BENARES,	HAKODATE,	KINGKITAO,	KHIVA,
NAGASAKI,	TEHERAN,	MEDINA,	SANA.
KESHO.			

* Numbers in list of mountains show height, in *thousands of feet*; in list of Rivers, lengths in *hundreds of miles*; in list of Cities and Towns, the population in thousands. For names differently spelled, see page 61.

A Procession in Japan, in honor of the Mikado

DESCRIPTIVE GEOGRAPHY.

1. ***ASIA*** is remarkable for its extent of surface, its high mountains, extensive plains and deserts, excessive climates, vast population, and historical antiquity.

2. ***The Himalayas,*** the highest mountains in the world, are south of the central part of Asia. Northward, the slope is gradual, while southward, it is very abrupt.

3. ***The great Plateau Region*** of Asia is in the center; the great plains are in the north, west, and east.

4. ***The Highest Mountain*** on the globe is Mount Everest, 29,000 feet above the level of the sea; the plateau of Thibet is from 10,000 to 15,000 feet high.

5. ***Nearly every Country is rich in*** wheat, rice and other grains, besides precious metals and stones.

6. ***Famous for the Manufacture*** of rich shawls, carpets, silks and velvets, are India, Persia and Turkey.

7. ***SIBERIA*** is composed of forests, steppes or prairies, marshes and fertile tracts, with fresh and salt lakes.

8. ***The Northern Third of its Surface*** is chiefly a region of salt steppes and endless swamps, where the soil is perpetually frozen to the depth of hundreds of feet. The surface is thawed only two or three months of the year.

9. ***The Rivers*** of Siberia are large, and their lower or northern courses are frozen the greater part of the year.

10. ***Siberia is rich*** in gold, precious stones and fur-bearing animals.

11. ***The Population*** is composed of Russian settlers, criminals and political exiles; besides the original inhabitants, who are a very degraded race, given to the lowest forms of idolatry.

12. ***CHINA,*** which forms about one-third of the Chinese Empire, has an area about equal to that portion of the United States which lies east of the Rocky Mountains, and a population ten times as large as that of the United States.

13. ***Its Surface*** is mountainous in the west, with an eastward slope to the Pacific Ocean.

14. ***Its great Source of Wealth*** is in the fertile soil of its extensive plains, river valleys and hills, which are highly cultivated. The lowlands yield two crops each year; the soil is enriched by a deposit of fine mud which is left by the floods.

15. ***The Agricultural Products*** of China and Japan are tea, rice, cotton, sugar, grain and fruits.

16. ***Among its Manufactures*** are silk goods and porcelain ware, the first inventors of which were the Chinese.

17. ***Inland Trade*** is facilitated by important rivers, canals, creeks and lakes.

18. ***The Minerals*** of China and Japan include gold, silver, copper, lead, iron and coal.

19. ***THIBET,*** an elevated region, belongs chiefly to China. It has a dry climate, except in summer, and abounds in precious metals.

20. ***THE JAPAN EMPIRE*** which is no larger than California, has a population nearly equal to that of the United States.

21. ***Its Climate*** is rainy, but milder than that of China, in the same latitude. The country suffers from hurricanes, earthquakes and volcanic eruptions.

22. ***The Government,*** like that of China, is a despotism. The Emperor of Japan is called the Mikado.

COMPARATIVE AREAS.
SQUARE MILES.

Russia in Asia, 5,600,000.

United States, 3,603,884.

Japan Empire, 149,399.

California, 188,981.

Afghanistan, 278,500.

Austria, 240,319.

Burmah, 192,500.

Spain, 195,000

The City of Bagdad, on the Tigris River, the principal city in the eastern part of Turkey in Asia. The Mosque of [illegible]

COMPARATIVE POPULATIONS.

Chinese Empire, 446,000,000.

United States, 40,000,000.

Japan Empire, 34,000,000.

Gt. Britain & Ire., 32,000,000.

Turkey in Asia, 16,050,000.

Spain, 16,535,000.

Ceylon, 2,300,000.

Chili, 2,000,000.

23. ***ASIATIC TURKEY*** contains some of the most celebrated places in the world. Palestine, or the Holy Land, is in the south-west; Mt. Ararat is on the northeastern border; Damascus is the oldest, and Jerusalem, the most renowned city in the world.

24. ***Smyrna*** is the largest city in Asiatic Turkey.

25. ***TURKESTAN*** is high in the east, with a western slope to the Caspian Sea. Its western section is sandy, salt and barren.

26. ***It is now*** chiefly under the dominion of Russia.

27. ***PERSIA*** consists mainly of a desert plateau, covered with sand, salt and nitre, where the heat of summer and the cold of winter are excessive. Its fertile section is in the northwest, near the Caspian Sea. The population comprises various Asiatic races.

28. ***The Head of the Government*** is called the Shah.

29. ***Many of the Inhabitants*** of Persia, Turkestan', Afghanistan', Beloochistan', Turkey in Asia and Arabia are Nomads, a wandering people, who rear sheep, goats, horses and camels.

30. ***ARABIA*** is a dry, hot and barren region, low along the coasts and high in the interior.

31. ***The Arabs*** comprise two classes; those who live in the cities and are engaged in agriculture, trade, etc., and the Bedouins, who live in tents and wander from place to place in search of pasture for their flocks and herds.

32. ***Agriculture*** is carried on by means of artificial irrigation. The leading products are coffee, grain, dates, gum Arabic and medicines.

33. ***The Largest City*** is Muscat. Mecca is celebrated as the birth-place of Mohammed (born 571, A.D.), the founder of Mohammedanism. Mocha is famous for its large export of excellent coffee.

34. ***INDIA*** extends from the Arabian to the China Sea, a distance from east to west, about equal to that of the United States.

35. ***It is Remarkable*** for its hot, moist climate, rich soil and the abundance of its vegetable and animal life.

36. ***The Forests, Valleys, Slopes*** and river banks contain valuable timber and a wonderful profusion of flowering vines, shrubs and trees.

37. ***The Leading Products*** are rice, wheat, cotton, coffee, opium, silk and tropical fruits.

38. ***The Wild Animals*** include the elephant, rhinoceros, lion, tiger, leopard, panther, bears and others. The forests teem with monkeys and beautiful birds.

39. ***The Useful Animals*** of Asia include the camel, horse, ox and donkey. The elephant and buffalo are also domesticated.

40. ***The Precious Metals*** and precious stones are abundant.

41. ***Hindoostan*** covers an area nearly equal to one-half that of the United States, with more than four times their population. It is mainly under the dominion of Great Britain. In the north, the cold of winter and the heat of summer are excessive; on the table-lands of the center and south, the climate is comparatively temperate; but, on the thickly settled plains, the heat, during the greater portion of the year, is intense.

42. ***AFGHANISTAN and BELOOCHISTAN*** consist of high plateaus and mountains. Their climates are excessive.

43. ***The Asiatic Governments*** are despotic; the religions, very diverse. Buddhism (bood'ism), an idolatrous worship, prevails in Thibet, China and Japan; Brahminism, also idolatrous, in India. The religion of Siberia is chiefly that of the Greek Church; that of Western and Southwestern Asia, Mohammedan, except Greeks and Armenians, who are Christians.

Showing the course of the Congo River, discovered by Stanley in 1877.

* Source of the Nile, discovered by Stanley, 1875.

SECTION OF AFRICA FROM CAPE VERD, EASTWARD, TO THE RED SEA.

EXERCISES ON THE MAP.

What natural division of land is AFRICA? By what waters is it embraced?

What prevents it from being an island? To what is Africa joined?

Where does Africa approach nearest to Europe?

What line is drawn east and west, nearly midway between the Mediterranean Sea and the Southern Ocean?

Is the greater part of Africa north, or south, of the Equator?

Near what parallel of latitude is Africa widest?

Does Africa become wider, or narrower, toward the south?

In what part is Sahara? What tropic passes through its center?

What desert in the southern part of Africa?

What tropic passes through it?

In how many zones does Africa lie? In which is its greater part?

What countries in the North Temperate? In the Torrid? In the South Temperate?

Are the mountains of Africa in the interior, or near the coasts?

What mountains in the north? With what coasts are they parallel?

What range of mountains in the east? South-east?

With what coast are they parallel?

What mountains in the southern part of Africa? With what are they parallel?

What ranges in the west? Near what coast are they?

On which side of Africa are the greatest elevations?

What river rises in the elevated regions of Africa?

What large lake is the source of the Nile River?

In what latitude is Lake Victoria Nyanza? What lakes south?

What lake in the central part of Africa?

How does the elevation of Lake Victoria Nyanza compare with that of Lake Tchad? (See Sectional View.)

What two high mountains near Lake Victoria Nyanza?

In what directions does the land slope which lies between the mountain ranges and the sea coast?

Name the rivers which flow into the Indian Ocean,—the Atlantic,—the Gulf of Guinea,—the Mediterranean Sea.

What rivers of Africa have deltas?

Name the Barbary States. By what bounded on the north? South?

Bound EGYPT. What is its capital? Its large river?

What towns on the Nile? On the Mediterranean Sea?

What town in the delta of the Nile? At the head of the Red Sea?

Bound NUBIA. Name its principal towns.

What two rivers unite in Nubia? From what directions do they flow?

Bound ABYSSINIA. What is its capital?

What divisions of Africa border on the Indian Ocean? On Mozambique Channel? On the Atlantic Ocean? On the Gulf of Guinea?

What large division south of Sahara?

Bound SOUDAN. Has Soudan any sea coast?

What other extensive region has no sea coast?

What lake near the center of Soudan?

What large river flows through its western part? What towns on the Niger? What towns in the north-east?

Bound SENEGAMBIA. What cape on the north-western coast?

What Republic south-east of Senegambia? British colony south?

What is the capital of LIBERIA? What are the divisions of Guinea?

Name the capitals of the kingdoms in Guinea. In Lower Guinea.

Bound CAPE COLONY. What is its capital?

Bound the GREAT DESERT. What countries partly covered by it?

What important oases in the Great Desert?

What is the capital of Madagascar? What capes on its coast?

What is the capital of Fezzan?

REVIEW.

COUNTRIES OR DIVISIONS.

Where are they? On what water do they border?

Morocco, or Marocco?	Guinea?	Liberia?	Congo?
Abyssinia?	Lower Guinea?	Zanguebar?	Soudan?
Senegambia?	Cape Colony?	Ashantee?	Egypt?
Orange Free State?	Hottentot?	Algeria?	Tunis?
Transvaal?	Caffraria?	Barca?	Nubia?
	Central Africa?	Benin?	Natal?
Mozambique?	Sierra Leone?	Dahomey?	Tripoli?

MOUNTAINS.

Where are they? In what directions do the ranges extend?

Lokinga?	Kilimandjaro?	Lupata?	Atlas?
Cameroon?	Ke'nia?	Snow?	Kong?

ISLANDS.

Where are they? By what waters are they surrounded?

Madeira Is.?	Madagascar?	Socotra?	St. Helena?
	Canary Is.?		Com'oro Is.?

GULFS AND SEAS.

Where are they? Into what waters do they open?

G. of Guinea?	G. of Aden?	G. of Sidra?	Red S.?

RIVERS.

Where do they rise? What courses do they take? Into what do they flow?

Nile?	Zambeze?	Tchadda?	Congo?
Niger?	Senegal?	Orange?	Coanza?

LAKES.

Where are they situated?

Victoria Nyanza?	Tanganika?	Nyassa?	Tchad?

CAPES.

Where are they? Into what waters do they project?

Good Hope?	Corrientes?	Palmas?	Bon?
Blanco?	Lopez?	Verd?	Ambro?
Guardafui?	Agulhas?	St. Lucia?	Frio?
		St. Mary?	

CITIES AND TOWNS.

Where are they? On or near what waters?

CAIRO,	COOMASSIE,	MONROVIA,	LOANGO,
TUNIS,	ST. SALVADOR,	MOURZOUK,	UJIJI,
MOROCCO,	TANANARIVOU,	TRIPOLI,	DERR,
ALGIERS,	MAGADOXO,	CAPE TOWN,	SUEZ,
ALEXANDRIA,	TIMBUCTOO,	GONDAR,	COBBE,
SACCATOO,	ABOMEY,	SIOUTT,	GONDOKORO

* Numbers in list of Mountains show heights, in thousands of feet; in list of Rivers, lengths in hundreds of miles; in list of Cities and Towns, population, in thousands.

Egyptian Ruins at Thebes

1. *AFRICA* is remarkable for its high surface, extensive deserts and hot climate.

2. *It is mainly in* the Torrid Zone. In latitude, its northern cape corresponds with Richmond (Virginia) and San Francisco; its southern, with Buenos Ayres.

3. *Its Mountain Chains* are almost parallel with the coast lines.

4. *The Coasts* are low and unhealthy, but the interior consists of high table lands.

5. *Its Great Desert* is Sahara; its principal rivers are the Nile, Niger, Zambese and Congo; its lakes are in Soudan and C. Africa.

6. *The Africans* include several distinct races. In the north, are Moors, Berbers, Arabs, Turks and Egyptians, all of whom belong to the Caucasian race, having straight hair; their complexion is dark, but not black. The southern half of Africa is the home of the Negro.

7. *The Productions of Africa* consist of wheat and other grains, cotton, sugar, coffee, tobacco, indigo, ivory, ebony, ostrich feathers, palm oil and tropical fruits.

8. *Inland Trade* is carried on by means of caravans.

9. *Celebrated for their Explorations* in Africa, are Barth, Du Chaillu, Grant, Speke, Baker and Livingstone and Stanley.

10. *The Animals* are numerous: they include the lion, elephant, rhinoceros, hippopotamus, leopard and giraffe. The gorilla is found in the Equatorial regions, and the crocodile in the rivers.

11. *EGYPT* corresponds in latitude with Florida: it extends from the Mediterranean Sea to the first Cataract of the Nile; it has the Red Sea on the east, and the desert on the west.

12. *It is Noted* for its antiquity, former greatness, gigantic pyramids, temples, obelisks, statues and tombs, constructed about 4,000 years ago.

13. *Its Climate* is dry and hot.

14. *The Fertility of its Soil* is wholly due to the annual rise of the water of the Nile, caused by periodical rains, which fall in torrents on the plateaus of Abyssinia and Central Africa. Without the Nile, Egypt would be a desert.

15. *In Egypt, the Nile is Lined* with high embankments, and the water is conducted in narrow channels to various parts of the valley, and allowed to cover the land, leaving a rich sediment.

16. *The Water Continues to Rise* at Cairo from June to September.

17. *The Valley* of the Nile varies in width, from 4 to 10 miles.

18. *The Southern Sources* of the Nile are several degrees south of the Equator.

19. *The Chief Exports* from Egypt, or through it from India, are silk and cotton.

20. *The Trees* of Egypt are the date-palm, acacia and tamarisk; there are no timber forests.

21. *The Useful Animals* are the horse, camel and donkey. Milk is obtained from the goat and the domestic buffalo.

22. *The Inhabitants* comprise Fellahs, Copts, Arabs and Turks; the Fellahs are the peasants and laborers, the Arabs are Bedouins, and the Turks are the ruling class. All are Mohammedans, except the Copts, who profess Christianity, and are the reputed descendants of the ancient Egyptians. Besides these, there are many Europeans and Jews. The language of the inhabitants is Arabic; their complexion, a brownish yellow.

23. *The Government* of Egypt, which is despotic and oppressive, is administered by the Khedive (king), who is subject to the Sultan of Turkey. Until recently, the ruler of Egypt was called the Viceroy.

24. *The History* includes six periods, the country being successively ruled by the Pharaohs, Ptolemies, Romans, Arabs and Turks.

25. *The Pyramid of Cheops* (ke'ops) is the largest; it is over 450 feet in height, and its base covers an area of 12 acres.

26. *Trade and Travel* between Europe and India are facilitated by the railroads between Alexandria, Cairo, and Suez, and by the Suez Canal, which connects the Red with the Mediterranean Sea.

27. *Cairo* is the capital and largest city. It is in the same latitude as New Orleans, and contains a population about equal to that of St. Louis and Kansas City combined.

28. *Alexandria*, next in size to Cairo, is the principal sea-port; it was founded by Alexander the Great, after whom it was named.

Populations of Countries.

Egypt, 4,915,000.

Sahara, 4,000,000.

Morocco, 2,750,000.

Algeria, 2,921,246.

Liberia, 720,000.

Cape Colony, 681,000.

Sierra Leone, 57,000.

Populations of Cities.

Cairo, 354,000.

Alexandria, 220,000.

Tunis, 129,000.

Morocco, 50,000.

Algiers, 58,000.

Fez, 100,000.

Mequinez, 62,000.

A Traveler and his Attendants, in Central Africa.

29. ***NUBIA*** has the same general characteristics as Egypt.

30. ***The Inhabitants*** are partly of Arab descent, but of a darker complexion; many are black, with Caucasian features. They cultivate the soil, and rear cattle, sheep and goats.

31. ***Nubia sends to Egypt,*** hides, ivory, ebony and ostrich feathers. It is subject to the Khedive of Egypt.

32. ***ABYSSINIA*** is a high table land, between 7,000 and 10,000 feet above the level of the sea, crowned with mountains, and connected with the great plateau of Southern Africa.

33. ***The Climate*** varies with the elevation, from the hot, pestilential coasts, to the limit of perpetual snow; on the high lands the climate is delightful. During the wet season, the rain falls in torrents.

34. ***The Products*** are grain, coffee, cotton, sugar, fruits, gums, and medicines.

35. ***The Domestic Animals*** are numerous.

36. ***The Population*** comprises various races—Ethiopians, Arabs, Jews, savage Gallas, and Negroes in a state of slavery.

37. ***The Abyssinian Religion*** is a branch of the Coptic Church; Mohammedanism, Judaism and Paganism also exist.

38. ***THE BARBARY STATES*** comprise Morocco, Algeria, Tunis and Tripoli; subject to Tripoli, are Barca and Fezzan.

39. ***The Principal Productions*** are Morocco leather, wool, grain, fruits, gum, olive oil, ostrich feathers and cabinet woods.

40. ***Cattle, Horses and Goats*** are numerous.

41. ***The Inhabitants*** are chiefly Mohammedans, and comprise Moors, Berbers, Arabs and Turks—all Caucasians, but of a dark complexion and straight hair. The Moors are descendants of Mohammedans who were expelled from Spain; they are of a dark yellow color. The people sometimes suffer greatly from famine, produced by drought or locusts.

42. ***Morocco*** is an empire, Algeria belongs to France, Tunis (ancient Carthage) and Tripoli are each governed by a Bey, who is subject to the Sultan of Turkey.

43. ***SENEGAMBIA*** is a region of abundant rains and luxuriant vegetation. It is named from its two rivers, the Senegal and Gambia, which abound in crocodiles and hippopotami.

44. ***Forests*** of valuable woods are extensive. Agriculture receives some attention.

45. ***The Inhabitants*** comprise a variety of tribes, each under its own chief. Some are civilized.

46. ***LIBERIA*** is a Negro republic, with an area about equal to that of Maryland.

47. ***SIERRA LEONE*** (se-er'ra la-o'na) is a British colony; so, also, are Cape Colony, Caffraria and Natal, in South Africa.

48. ***Upper Guinea*** and Lower Guinea are inhabited by numerous Negro tribes, mostly pagans, some of whom are engaged in agriculture and manufactures, while others are addicted to most barbarous practices; on the coasts are British, Dutch and Portuguese settlements.

49. ***MOZAMBIQUE*** is claimed by the Portuguese.

50. ***On the Western Side*** of Zanguebar are the highest mountains of Africa, Kenia and Kilimandjaro, each about 20,000.

51. ***MADAGASCAR*** is a hereditary monarchy, covering an area larger than that of France. The inhabitants are rapidly advancing in civilization, and are chiefly engaged in rearing cattle.

EXERCISES ON THE MAP.

What are the divisions of **OCEANICA**?

Which contains the largest islands?

Micronesia signifies small islands; Polynesia, many islands.

Mention the largest island. What division is between Australia and Asia? Name the largest island in Malaysia.

Through which does the Equator pass? What towns in Sumatra? In Borneo?

What is the capital of Java? What Island east of Borneo?

What group of islands in the northern part of Malaysia? In the eastern?

Name the divisions of Australia. What cities in the south-east?

What capes on the north-east? South-east? North-west? South-west? Where is the gold region of Australia?

What separates Van Dieman's Land from Australia?

REVIEW.

ISLANDS.

Where are they? By what waters are they surrounded? In what part of what division are they?

Java? Mendana Arch.? Ladrone Is.?
Sandwich Is.? New Zealand? Caroline Is.?
New Guinea? Friendly Is.? New Ireland?
Borneo? Magellan's Arch.? Feejee Is.?
Australia? [illegible] Society Is.? Spice Is.?
Tasmania? Luzon?
Philippine Is.? [illegible] Sumatra? [illegible] Celebes? [illegible]
Hawaii? [illegible] Central Arch.? [illegible] Marquesas? [illegible]

CITIES AND TOWNS.

Where are they? On or near what waters?

MELBOURNE, BENCOOLEN, BATAVIA, AUCKLAND,
ADELAIDE, ACHEEN, BORNEO, SYDNEY.

STRAITS.

Between what lands are they? What waters do they connect?

Malacca? Torres? Sunda?
Macassar? Cook's? Bass?

SEAS, GULFS, AND BAYS.

Where are they? Into what waters do they open?

G. of Carpentaria? Coral Sea? Botany Bay?
G. of Cambridge? China Sea? Spencer's Gulf?
Celebes Sea? Java Sea? G. of Siam?

1. ***OCEANICA*** comprises nearly all the islands of the Pacific Ocean.

2. ***The most Important*** divisions are Australasia and Malaysia.

3. ***The Largest Islands*** in Australasia are Australia, Pap'ua, or New Guinea, New Zealand and Tasmania.

4. ***AUSTRALIA*** is the largest island in the world; having an area equal to five-sixths that of Europe.

5. ***It lies*** partly in the Torrid, and partly in the Temperate Zone, the Tropic of Capricorn passing nearly through its center.

6. ***Its Surface*** is generally flat, with a border of hills or mountains near the coasts. There are no large rivers which lead into the interior. The largest river—the Murray—is in the southeast.

7. ***The Climate*** of the northern or Tropical portion is hot and dry; while in the south, it is delightful during eight or nine months of the year.

8. ***Being in the Southern Hemisphere***, the north wind is hot, and the south wind, cold; the hottest months are December, January and February; the coldest, July, August and September.

9. ***The North Wind*** is, at times, like the blast from a furnace, the mercury rising to 130° and 140° Fahr.

10. ***Australia belongs*** to Great Britain, and comprises the Colonies of Victoria, New South Wales, Queensland, South Australia, West Australia, and the territory of North Australia. The settlements are on or near the coasts, and especially in the southeast.

11. ***Victoria*** is the richest and most populous colony, and is remarkable for the beauty of its scenery.

12. ***The Chief Products*** of Australia are gold, copper and wool; gold is obtained from Victoria and New South Wales; copper and iron, from South Australia; wool and the useful metals from nearly all the colonies. Diamonds have been recently found in New South Wales.

13. ***The most Productive Soil*** is in the east and southeast: the products being wheat, oats and other grains, potatoes, sugar, tobacco, cotton and tropical fruits.

14. ***The Interior*** has recently been found to contain extensive regions which are well watered and adapted to pasture and agriculture.

15. ***The Wild Animals*** are not numerous; the most important is the kangaroo, which moves rapidly by a succession of bounds. The emu, a bird resembling an ostrich, stands about six feet high.

16. ***The Domestic Animals***, originally introduced from Europe, are now counted by millions; sheep are especially numerous. Llamas and alpacas, valuable fleece-bearing animals, have been introduced from South America.

17. ***The Original Inhabitants*** are short and stout, with small heads, flat noses, thick, protruding lips, and long, coarse hair. Their complexion is black or dark brown. They live in holes or hovels covered with the bark or branches of trees, and are wholly or nearly destitute of clothing. Their number is rapidly diminishing.

18. ***The Population***, exclusive of the natives, consists of British settlers and their descendants, with some Germans and Chinese.

19. ***The First White Persons*** in Australia were convicts from England. New South Wales was a penal colony from 1788 to 1840. Most of the prisoners, upon the expiration of their sentences, remained, and merged in the general population, which rapidly increased after the discovery of gold, in 1851.

20. ***The Principal Cities*** are Melbourne, Sydney and Adelaide.

21. ***Pap'ua or New Guinea*** is mountainous, and covered with forests; the sago, camphor and cocoa-nut trees are plentiful; also rice, corn and spices.

22. ***Its Inhabitants*** are savages, resembling the native Australians.

A New Zealand Chief.

23. ***New Zealand***, a British Colony, covers an area about twice as large as that of the New England States.

24. ***The Peculiarities*** are its mountains, which rise above the snow limit, its volcanoes and earthquakes, extensive forests, rich plains and valleys, and heavy rains in mid-winter (July).

25. ***The Population*** is chiefly composed of white colonists. The aborigines, are Malays, who were formerly fierce cannibals, but they are now more or less civilized; they are tall, active and well-formed, have prominent features, straight hair, and a light brown complexion (see page 46, paragraphs 43, 44 and 45).

26. ***The Leading Occupations*** are agriculture and grazing.

27. ***MALAYSIA*** is noted for its hot, moist climate, luxuriant vegetation, numerous volcanoes and frequent earthquakes.

28. ***Its Products*** include bamboos, rattans, teak, ebony, gutta percha and sandal-wood, besides cloves, nutmegs, pepper, ginger, cinnamon, rice, cotton, tobacco, coffee, sugar and fruits.

29. ***The Animals*** are the elephant, rhinoceros, tiger, panther, monkeys and orang-outangs. Birds of Paradise, parrots and other birds of beautiful plumage, are here in great abundance.

30. ***The Inhabitants*** are Malays, Papuan negroes, Dutch, English and Chinese.

31. ***THE SANDWICH ISLANDS*** are important as a central station for ships trading between the United States and Asia.

32. ***They are Remarkable*** for volcanoes and earthquakes.

33. ***The Climate*** is mild and healthful. Tropical plants grow on the low ground; rice, wheat, potatoes, etc., on the higher, or temperate regions.

34. ***The Largest*** of the Sandwich Islands is Hawaii (hah wi'e), on which is situated the celebrated volcano, Mauna Loa.

35. ***The Inhabitants***, chiefly of the Malay race, are rapidly advancing in civilization.

36. ***The Government*** is a kingdom; the capital and chief city is Honolulu.

COMPARATIVE EXTENT AND LATITUDES.

EXERCISES ON THE MAPS.

The Names and Colors on the right and left hand margins of the Maps indicate the latitudes of countries, islands, etc., and their extent from north to south, regulated to the scale of each map. Those on the upper and lower margins show their comparative extent from east to west.

Map of North America.—(Page 52.)

Through what parts of North America does the Arctic Circle pass? The Tropic of Cancer?

Between what parallels of latitude do the United States lie? *Ans. Between the parallels of 25° and 49° north latitude.*

What parts of North America lie between the parallels of 20° and 40° north latitude? Between those of 40° and 50°? Between those of 50° and 60°?

Refer to the right-hand margin and tell what countries on the eastern side of the Atlantic Ocean lie between the parallels of 50° and 60° north latitude?

What country of Europe is directly east of the northern part of Labrador? Of the southern part? East of the British Provinces?

At what country would you arrive by sailing eastwardly across the Atlantic from Cape Race? From Cape May? From Cape Hatteras? From Cape Sable, of Florida?

What difference exists between the climate of the eastern coast of the United States and those of the western Coasts of Europe and Africa? Why? (*See page 96, paragraphs* 11, 14, *and* 24.

What part of the Old World is between the same parallels as the West Indies? (*See both margins.*) As the Caribbe'an Sea?

What country in Asia lies directly west of the western coast of the United States?

What two empires west of the United States?

What parts of Asia, North America, and Europe lie between the parallels of 40° and 60° north latitude? Between those of 20° and 40°? Below that of 20°?

What countries of Europe lie in a line eastward from the Great Lakes of North America?

What countries of Europe are further north than Canada?

Which is further north, Newfoundland or Ireland? Quebec or London?

What cities of the United States are in the same latitude as Spain? Morocco?

What part of what country is eastward from the northern boundary of Maine?

What parts of the Old World are opposite the eastern coast of the United States? The western coast?

Map of the United States.—(Page 58.)

Mention all the States that are in the same latitude as any part of France? Spain? Morocco?

What part of what country is in a line eastward from your State?

What States are directly east from the center of your State? West?

Mention the principal cities of the United States that are in the same, or nearly the same, latitude as the capital of your State.

Mention those of the Old World that are in that latitude.

Map of the Eastern States.—(Page 60.)

What Western State is here shown to be west of Maine? What country in Europe east of the northern part of Maine? Of its southern part?

What cities in Maine lie between the same parallels of latitude as Northern Italy? What cities in New Hampshire and Vermont lie between those parallels?

What is the latitude of the northern boundary of Vermont?

Which of these States lie between the same parallels as Iowa? As Spain?

What is the latitude of the most southern part of Connecticut?

Map of the Middle States.—(Page 62.)

What countries in Europe lie east of the Middle States?

What three Western States west?

What part of what country in Europe is directly east of Northern New York? Of Central New York? Of the south-eastern part of the State?

Which of the Middle States lie between the same parallels of latitude as Portugal?

What is the latitude of the northern boundary of Pennsylvania? Of Portugal?

Map of the Southern States.—(Page 64.)

What parts of the Old World lie between the same parallels as the Southern States?

Which of these States extend south of the parallel of 30°?

What is the latitude of the northern boundaries of North Carolina and Tennessee?

What cities on or near the parallel of 30°? Of 32°? Of 36°?

Map of the Western States.—(Page 66.)

What Pacific State lies west of Michigan and Wisconsin?

What Western States lie between the same parallels as Nevada?

What country in Europe is in the same latitude as the northern part of Minnesota? The northern part of Michigan?

Which of these States lie between the same parallels of latitude as the northern half of Italy? The southern half?

Which are in the same latitude as Sicily?

Map of the Western Territories.—(Page 68.)

What is the latitude of the northern boundary of Montana?

What capital city is near the center of the Union? *Ans. Topeka.*

Name the States and cities in a line west of Topeka. East.

What Territories are in a line westward from Southern France?

What States and Territories lie wholly or partly between the same parallels as Spain? Morocco?

What Asiatic islands west of Oregon and Washington?

By sailing westwardly from San Francisco, at what country would you arrive?

Map of South America.—(Page 80.)

What countries lie within the same parallels as Australia?

Over how many degrees of latitude does Australia extend from north to south?

What large island of Malaysia, and what countries of South America are crossed by the Equator?

What country is about the same in extent from east to west as Brazil? (*See upper margin.*) What island? (*See lower margin.*)

Map of the British Provinces.—(Page 56.)

What country in Europe lies directly east of the British Provinces?

What Pacific State west of the southern part of Canada? What Territory west of the northern part?

Is any part of Canada further south than Boston? Albany?

What capital cities in the United States are in the same, or nearly the same, latitude as Kingston?

What city in Canada is in nearly the same latitude as Concord? Portland?

Index to Contents

ARRANGED AS A

GENERAL REVIEW OF THE PHYSICAL GEOGRAPHY.

ASTRONOMICAL GEOGRAPHY.

(THE WORDS IN BLACK TYPE SUGGEST THE QUESTIONS.)

1. ***Astronomical Geography treats*** of the form, size, and motions of the earth; its relations to the Sun, Moon, and other heavenly bodies; its seasons, latitudes, and longitudes.

2. ***The Earth*** is one of a family of heavenly bodies which revolve around the Sun.

3. ***The bodies which revolve around the Sun are distributed*** into three classes; Planets, Asteroids, and Comets.

4. ***These bodies, together with the Sun, constitute*** the Solar System.

5. The ***Solar System*** is but a small portion of the Universe.

6. ***The Sun*** is a luminous body, because it shines by its own light. The planets are opaque (dark) bodies.

7. ***The Earth, Moon, and other planets receive from the Sun*** light and heat.

8. ***The names of the principal planets,*** according to their size, are Jupiter, Saturn, Neptune, Ura'nus, the Earth, Venus, Mars, and Mercury.

9. ***Their names according to their distances*** from the Sun, are Mercury, Venus, the Earth, Mars, Jupiter, Saturn, Uranus, and Neptune.

10. ***The form of the Earth*** is that of a sphere, slightly flattened at the Poles. (*See illustration on page 9.*)

11. ***A Sphere or Globe*** is a round body whose surface, in every part, is equally distant from its center.

12. ***A Hemisphere*** is half a sphere or globe.

13. ***The Diameter of a Sphere*** is a straight line passing through its center, and terminated at both ends by the surface.

14. ***The Diameter of the Earth*** is nearly 8,000 miles.

Its diameter at the Equator is 7,925 miles, but from Pole to Pole it is 26 miles less.

15. ***The Circumference of a Sphere*** is the distance around it.

16. ***The Circumference of the Earth*** is nearly 25,000 miles.

17. ***The Axis of a Sphere*** is the line or diameter on which the sphere revolves.

18. ***The Poles of the Earth,*** or of any sphere, are the extremities of its axis, or the two points where the axis meets the surface.

19. ***The Sun shines upon*** one half of the earth's surface at any one time; so that one hemisphere has day while the opposite hemisphere has night.

20. ***The succession of Day and Night*** is caused by the revolution of the earth on its axis, which it performs every 24 hours.

21. ***The rate of Motion*** of the equatorial parts is 1,000 miles every hour, but it diminishes toward the Poles.

The Axis does not revolve, neither do the Poles.

22. ***Localities on the Earth's surface are determined*** and described by means of imaginary lines or circles.

23. ***Great Circles*** are those which divide the Earth into two equal parts.

24. ***Small Circles*** are those which divide the Earth into two unequal parts.

25. ***The principal Great Circles*** are the Equator, Ecliptic, and Meridians.

26. ***The principal Small Circles*** are the two Tropics and the two Polar Circles.

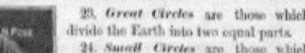

27. ***The Equator divides the Earth into*** Northern and Southern Hemispheres. *It is midway between the Poles.*

28. ***Meridians pass*** from Pole to Pole, crossing the Equator at right angles.

29. ***Meridians divide the Earth*** into Eastern and Western Hemispheres.

30. ***Latitude*** is distance northward or southward from the Equator, measured on a Meridian.

31. ***Longitude*** is distance eastward or westward from a certain Meridian, measured on the Equator.

32. ***Latitude and Longitude are reckoned*** in degrees, minutes, and seconds, which are known by the signs (°), ('), (").

The City Hall of New York is in lat. 40° 42' 43" (read 40 degrees, 42 minutes, and 43 seconds). A degree contains 60 minutes, and a minute 60 seconds.

33. ***A Degree*** is one 360th part of a circle; it varies in length according to the size of the circle.

34. ***The length of a degree*** on a Great Circle of the Earth is about 69½ statute miles, or 60 geographical miles.

A statute mile contains 5,280 feet, and a geographical mile, 6,075 feet.

35. ***The parts of the Earth farthest from the Equator are*** the Poles, whose latitude is 90°.

36. ***Longitude is usually reckoned,*** on our maps and globes, from the Meridian of Greenwich, near London, and from the Meridian of Washington.

37. ***The greatest Longitude*** a place can have is 180°—half way round the globe.

38. ***Refer to the Map*** on pages 52 and 53, and state the Latitude of Philadelphia; of New Orleans; of Columbus; of Nashville; of San Francisco; of Savannah.

39. ***What is the Longitude of each,*** from Greenwich, and from Washington?

40. ***Refer to the Map on page 72,*** and state the Latitude of Naples; of Venice; of Lucerne; of Athens; of Constantinople; of Paris; of Frankfort; of Hamburg; of London; of Liverpool; of Dublin.

41. ***What is the Longitude of*** London? of Dublin? of Geneva? of Rome? of Vienna?

42. ***The Ecliptic*** is the path in which the Earth revolves around the Sun. In Geography, the Ecliptic is a great circle on the terrestrial globe which is always in the plane of the earth's orbit.

43. ***The Equator and Ecliptic cross each other*** at an angle of 23½°.

44. ***The Sensible Horizon*** is the Small Circle which bounds our view of the Earth's surface. *Its circumference* is the line in which the Earth and Skies appear to meet; *spectators in different localities have* different horizons. *In the middle of the horizon* is the spectator. The higher the elevation on which the spectator stands, the greater is the sensible horizon. A person at sea, standing on the level of the surface, would see three miles in every direction. *The diameter* of his sensible horizon would be six miles. (*See page 9, illustration, and paragraphs 1 to 5.*)

45. ***The Rational Horizon*** is the Great Circle which is parallel to the Sensible Horizon; it divides the Earth into *upper* and *lower hemispheres.*

46. *Parallels of Latitude* are small circles parallel to the Equator.

47. *The Tropics* are those parallels which pass through the two points of the Ecliptic farthest from the Equator.

48. *The Tropic in the Northern Hemisphere is called* the Tropic of Cancer. *That in the Southern Hemisphere*, the Tropic of Capricorn.

49. *The Distance of the Tropics from the Equator* is 23°.

50. *The Axis of the Earth* is not perpendicular to the plane of the Earth's orbit.

51. *The Distance from the Poles to the Extremities of a Diameter* which is perpendicular to the Ecliptic is 23½°; through these two extremities two parallels of latitude are drawn; that around the North Pole is called the Arctic Circle or North Polar Circle, and that around the South Pole, the Antarctic, or South Polar, Circle. (*See illustration at the top of the page.*)

52. *The Tropics and Polar Circles divide* the Earth's surface into five great Belts or Zones. (*See map on page 81.*)

53. THE ZONES AND THEIR EXTENT FROM NORTH TO SOUTH.

North Frigid....	From the North Pole to the Arctic Circle............	23½°
North Temperate	Arctic Circle to the Tropic of Cancer.......	43°
Torrid..........	Tropic of Cancer to the Tropic of Capricorn..	47°
South Temperate "	Tropic of Capricorn to the Antarctic Circle..	43°
South Frigid... "	Antarctic Circle to the South Pole.........	23½°
Total, from Pole to Pole................................		180°

54. *Within the Torrid Zone the Heat* is extreme, because the Sun's rays fall *directly* upon the surface.

55. *The Cold of our Winter* is not known, except at high elevations. (*See page 39, paragraphs 62–66.*)

56. *The Days and Nights* on and near the Equator are equal throughout the year. Leaving the Equator, *their inequality increases* with the latitude.

57. *The Sun is Vertical to* the inhabitants of the Torrid Zone at certain times during the year. (*Read page 45, par. 16, 17, 22, and 23.*)

58. *The Sun is Vertical, or in the Zenith, when* it is perpendicularly over the head.

59. *Within the Frigid Zones the Cold* is extreme, because the Sun's rays fall very *obliquely* upon the surface.

> *The Longest Days* in Summer and the *Longest Nights* in Winter are in proportion to the latitudes,—from 24 hours on the Polar Circles to 6 months at the Poles.
>
> *The Sun is never Vertical* to any of the inhabitants of the Frigid Zones.

60. *Within the Temperate Zones the Heat* is less than that in the Torrid Zone, and the cold is less than that in the Frigid Zones.

> *The Longest Days* in Summer and the *Longest Nights* in Winter vary from 13½ hours on the Tropics to 24 hours on the Polar Circles.
>
> *The Sun is Vertical* once a year—midsummer—to the inhabitants on the Tropics.

61. *The Change of Seasons depends* upon the annual revolution of the earth in the same plane, the inclination of its axis, and the leaning of the axis always in the same direction.

62. *The North Pole leans toward the Sun* in the latter part of June; then it is Summer in the Northern and Winter in the Southern Hemisphere. (*See illustration above.*)

> *The Northern Hemisphere has* long days and short nights, while the *Southern Hemisphere has* short days and long nights.
>
> *The Whole of the North Frigid Zone has* day, while the South Frigid has night.
>
> *The Sun is Vertical* to the inhabitants on the Tropic of Cancer.

63. *The North Pole leans from the Sun*, in the latter part of December; then it is Summer in the Southern and Winter in the Northern Hemisphere; the *Southern Hemisphere has* long days and short nights, while the *Northern* has short days and long nights.

> *The Whole of the South Frigid Zone has* day, while the *North Frigid has* night.
>
> *The Sun is Vertical* to the inhabitants on the Tropic of Capricorn.

64. *On the 23d of March*, neither the North nor the South Pole leans toward the Sun. (*In the illustration above, the pupil must imagine the Earth to have moved around behind the Sun*). Then it is *Spring* in the Northern Hemisphere while it is *Autumn* in the Southern; *the Sun is vertical* to the inhabitants on and near the Equator, and the line of separation between the dark and the illuminated side of the Earth passes through the Poles.

65. *On the 21st of June*, the position of the Earth is as represented in the picture; three months afterward, or on the *23d of September*, the Earth's position would be sidewise, as in March. (*In the picture imagine the Earth to have moved toward you, and to be immediately in front of the Sun, about two inches from the page.*)

66. On the 23d of September it is Autumn in the Northern, and Spring in the Southern Hemisphere,—12 hours day and 12 hours night, in all the Zones; the Sun vertical to the inhabitants on the Equator; the days and nights are everywhere equal.

GENERAL REVIEW.

MOUNTAINS.

Where are they? In what directions do the ranges extend?

Mt. St. Elias?	Spanish Peak?
Mt. Washington?	Erze Gebirge?
Mt. Mitchell?	Illimani?
Himalaya?	Adirondack?
Moravian?	Atlas?
Mt. Elboorz?	Western Ghauts?
Rocky?	Pike's Peak?
Mt. Brown?	Cumberland?
White Mts.?	Antuco?
Cameroon?	Blue Ridge?
Hindoo Coosh?	Kong?
Bohemian?	Caucasus?
Mt. Blanc?	Fremont's Peak?
Coast?	Catskill?
Kilimandjaro?	Antisana?
Kuen Lun?	Mt. Hood?
Cevennes?	Altai?
Mt. Fairweather?	Highlands?
Green Mts.?	Pacaraima?
Kenia?	Taurus?
Ararat?	Alps?
Auvergne?	Acaray?
Sierra Madre?	Carpathian?
Brazilian Andes?	Atacama?
Alleghany?	Ural?
Cotopaxi?	Mt. Etna?
Mt. Hooker?	Andes?
Lupata?	Cantabrian?
Stanovoy?	Geral?
Sierra Morena?	Apennines?
Fremont's Peak?	Arequipa?
Chimborazo?	Mt. Vesuvius?
Hecla?	Pichincha?
Snow?	Mt. Hecla?
Eastern Ghauts?	Popocatepetl?
Riesen Gebirge?	Sierra Nevada?
Cascade?	Scandinavian?
Aconcagua?	Pyrenees?
Coast Range?	Long's Peak?

ISLANDS.

Where are they? By what waters are they surrounded?

Newfoundland?	San Salvador?
Sardinia?	Mendana Arch.?
Saghalien?	Baring?
Jamaica?	West Indies?
Java?	Minorca?
British Is.?	Hainan?
Madeira Is.?	Antigua?
Orkney Is.?	New Zealand?
Vancouver's?	Bahamas?
Negropont?	Zante?
Nova Zembla?	Nicobar Is.?
Barbadoes?	Porto Rico?
Sandwich Is.?	Friendly Is.?
Loffoden Is.?	Bermudas?
Madagascar?	Formentera?
Hebrides?	Hondo?
Southampton?	I. of Pines?
Rhodes?	Society Is.?
Kurile Is.?	Melville?
Skye?	Abaco?
Magellan Arch.?	Central Arch.?
Louisiade Arch.?	Lewis?
Trinidad?	Jersey?
New Guinea?	Kiusiu?
Faroe Is.?	Andros?
Canary Is.?	Tasmania?
Anglesea?	Anticosti?
Queen Charlotte's?	Lipari Is.?
Ionian Is.?	Formosa?
Laccadive Is.?	Turks?
Guadaloupe?	Cuba?
Borneo?	Corsica
Shetland Is.?	Shikoku?
Socotra?	St. Thomas?
Wight?	Caroline Is.?
Cape Breton?	Iceland?
Scilly Is.?	Ivica?
Maldive Is.?	Japan Is.?
Martinique?	Santa Cruz?
Australia?	New Ireland?
Cyprus?	Disco?
St. Helena?	Corfu?
Man?	Yeso?
Greenland?	Feejee Is.?
Balearic Is.?	Sitka?
Ladrone Is.?	Majorca?
New Providence?	Luzon?
Hawaii?	Spice Is.?
Candia?	Hayti?
Comoro Is.?	Cephalonia
Philippine Is?	Ceylon?
North Georgian?	Celebes?
Sicily?	Guernsey?
Sumatra?	Marquesas?

CAPES.

Where are they? Into what waters do they project?

Hatteras?	East?
Corrientes?	St. Blas?
Land's End?	Finisterre?
St. Antonio?	St. Roque?
Northeast?	St. Martin?
Canaveral?	Cambodia?
Spartivento?	Florida?
Good Hope?	North?
Mendocino?	Agulhas?
St. Francisco?	Flattery?
La Hogue?	Corso?
Comorin?	Prince of Wales?
Haitona?	Fear?
Matapan?	Clear?
Blanco?	Palmas?
Farewell?	Sable?
Gallinas?	Horn?
Gracios?	Ortegal?
Lopatka?	Verd?
Lookout?	Race?
St. Vincent?	Bon?
Guardafui?	May?
St. Lucas?	Frio?
St. Lorenzo?	Cod?
Palmo?	Icy?
Negrais?	Rozo?
Burica?	St. Mary?

SEAS, GULFS, BAYS, &c.

Where are they? Into what waters do they open?

St. Lawrence?	James' B.?
Ionian Sea?	Zuyder Zee?
California?	Darien?
Appalachee?	Tampa?
The Wash?	Loch Linnhe?
Arabian S.?	Okhotsk S.?
Mediterranean?	Archipelago?
Narragansett?	Casco?
G. of Guinea?	G. of Siam?
G. of Carpentaria?	Ungava B.?
B. of Honduras?	Onslow?
G. of Lepanto?	Firth of Tay?
Campeachy?	Yellow S.?
Albemarle?	Baltic S.?
Firth of Forth?	Cape Cod?
Caspian S.?	Frobisher's B.?
Adriatic S.?	Pamlico?
Penobscot?	Galway?
G. of Aden?	Black S.?
G. of Cambridge?	Fundy?
B. of Guatemala?	Chan. of Yucatan?
G. of Genoa?	Murray Firth?
Tehuantepec?	Red S.?
Roanoke?	G. of Finland?
Donegal?	Fox Chan.?
Persian G.?	Kara S.?
B. of Biscay?	White S.?
Frenchman's?	Lancaster Sd.?
G. of Sidra?	Aral S.?
Coral S.?	G. of Lyons?
Hudson B.?	Norton Sd.?
G. of Salonica?	Eastern S.?
Mosquito?	S. of Marmora?
Raleigh?	G. of Mexico?
Pentland Firth?	Celebes S.?
G. of Cambay?	G. of Onega?
G. of Bothnia?	B. of Bengal?
Long Island Sd.?	North S.?
Java S.?	China S.?
Baffin B.?	S. of Azov?
G. of Dantzic?	G. of Siam?
Panama?	G. of Taranto?
Mobile?	Japan S.?
Loch Foyle?	G. of Riga?
Kamtschatka S.?	G. of Cutch?
Caspian S.?	S. of Yeso?
Buzzard's?	G. of Tonquin?
Botany?	

RIVERS.

Where do they rise? In what directions do they flow, and into what waters?

Mississippi?	Coanza?
Thames?	Brazos?
Volga?	Ucayali?
Wabash?	Congaree?
Meuse?	St. Francis?
Susquehanna?	Kentucky?
Lena?	Boyne?
Trent?	Dwina?
Missouri?	Sangamon?
Red?	Elbe?
Nile?	Alleghany?
Rio Grande?	Irrawaddy?
St. Francisco?	Grand?
Tombigby?	Appalachicola?
Penobscot?	Nelson?
Tennessee?	Orinoco?
Shannon?	Pascagoula?
Danube?	Otter Creek?
Des Moines?	Sandusky?
Adige?	Foyle?
St. Lawrence?	Garonne?
Hoang Ho?	Big Sandy?
St. Francis?	Pruth?
Green?	Genesee?
Niger?	Yenisei?
Athabasca?	Platte?
Tunguragua?	Tallapoosa?
Cape Fear?	Severn?
Kennebec?	La Plata?
Cumberland?	Suwanee?
Severn?	Blackstone?
Ural?	Grand (Mich.)?
Detroit?	Avon?
Monongahela?	Douro?
Cambodia?	Big Sioux?
St. Maurice?	Po?
Brazos?	Juniata?
White?	Obi?
Zambeze?	Lewis?
Mackenzie?	Big Black?
Magdalena?	Platte?
Mobile?	Parana?
Androscoggin?	Yazoo?
Wisconsin?	Sorel?
Mersey?	Grand (Mo.)?
Vistula?	Drave?
Illinois?	Little Sioux?
Spree?	Iser?
Shenandoah?	St. Regis?
Irish?	Tigris?
Saguenay?	Canadian?
Salmon?	Ocmulgee?
Senegal?	Uruguay?
Colorado?	St. John's?
Pilcomayo?	Onion?
Ogeechee?	Iowa?
Merrimac?	Guadiana?
Miami?	Osage?
Ouse?	Dniester?
Dnieper?	Pamunky?
Kalamazoo?	Indus?
Weser?	Yellow Stone?
Oswegatchie?	Pamlico?
Brahmapootra?	Ohio?
Chaudiere?	Amazon?
Columbia?	Edisto?
Chattahoochee?	Maumee?
Tchadda?	Tagus?
St. Lawrence?	Licking?
Paraguay?	Saranac?
Altamaha?	Amoor?
Connecticut?	Clarke's?
Muskingum?	Yadkin?
Humber?	Madeira?
Petchora?	Oconee?

PRONUNCIATION OF GEOGRAPHICAL NAMES.

A.

B.

C.

D.

E.

F.

G.

H.

I.

J.

K.

L.

M.

N.

O.

P.

Q.

R.

S.

T.

U.

V.

W.

Y.

Z.

MEANING OF GEOGRAPHICAL NAMES.

GEOGRAPHY

A View of the Yo Semite Valley, in the Sierra Nevada, looking up the Valley (E by N. E.). On the right or south, is the Bridal Veil Fall (630 feet); on the left, El Capitan, a perpendicular cliff (3300 feet).

OF THE

PACIFIC SLOPE.

OREGON
IDAHO
NEVADA
HUMBOLDT
LINCOLN
LOWER CALIFORNIA
PRINCIPAL PRODUCTS
Scale of Miles
SECTION ACROSS THE STATES OF CALIFORNIA & NEVADA
PENNSYLVANIA
NORTH CAROLINA
SOUTH CAROLINA

CALIFORNIA AND NEVADA.

EXERCISES ON THE MAP.

CALIFORNIA,—what parallel of latitude on its northern boundary? On or near its southern boundary? What river forms part of its boundary?

What is the length of California from north to south? (*Apply the scale of miles.*) What is its breadth? What high mountain range in the eastern part of the State? What range in the western part?

What two large rivers drain the great valley between these ranges?

Which flows from the north? From the south?

What tributaries has the Sacramento? The San Joaquin?

What large river in the north-western part of California? What lakes are drained by the Klamath River? What river empties into Monterey Bay?

What is the largest lake in California? What rivers empty into it?

What is the largest city in California? (*Ans.—San Francisco.*) On what bay is it situated?

What bays on the coast of California, north of San Francisco? South?

(*The teacher will adapt the following exercises to the State or Territory in which he resides.*)

In what county do you live? In what part of the State is it? Has it any sea coast? Any boundary river? Does any river run through it?

By what counties is your county surrounded?

What are the northern counties of your State? The southern? The central?

What is the capital of the State? How is it situated? What is the county town of your county? Of each of the surrounding counties?

Draw an outline of your State, beginning at the north-west corner, and proceed easterly, thence southerly, and so on. Insert, in order, the mountains, the rivers, the lakes, and the bays, with their names. Write your name on your paper or slate, and after your drawing has been examined by the teacher, proceed with the following lessons: Insert the counties and their county towns; next, insert the other cities and towns, then the islands, capes, &c., and complete the map.

NEVADA,—what parallel of latitude forms its northern boundary?

What State and Territory north of Nevada? What two Territories east?

What State west? What mountain chains in Nevada?

In what part of the State are they? In what direction do they extend?

What river on its south-eastern boundary?

What tributary of the Colorado flows through the south-eastern part of Nevada? What is the principal river in Nevada? Into what does it empty?

What tributaries has the Humboldt from the south? From the north?

What railroad runs along the Humboldt Valley?

In what part of the State are most of its lakes? Mention their names?

Have they any outlets to the ocean?

What water communication has Nevada with the ocean?

What counties border on California? On Oregon? On Idaho? On Utah? On the Colorado River? What counties do not extend to any part of the border? Through what counties does the Pacific Railroad pass?

REVIEW.

CITIES AND TOWNS.

In what part of the State are they situated? On or near what waters? In what counties are they? Which are county towns?

CALIFORNIA.

SAN FRANCISCO, SACRAMENTO, SAN JOSE, MARYSVILLE, STOCKTON, NEVADA CITY, CRESCENT CITY, ORLEANS BAR, YREKA, TRINIDAD, EUREKA, HUMBOLDT CITY, WEAVERVILLE, CANON CITY, SHASTA CITY, SUSANVILLE, QUINCY, RED BLUFF, GROVE CITY, OROVILLE,

DOWNIEVILLE, COLUSA, UKIAH CITY, YUBA CITY, LAKEPORT, WOODLAND, AUBURN, PLACERVILLE, COLOMA, SANTA ROSA, NAPA, FAIRFIELD, BENICIA, PETALUMA, JACKSON, SILVER MT. CITY, VALLEJO, BRIDGEPORT, MONOVILLE, SAN ANDREAS,

SAN RAFAEL, MARTINEZ, SONORA, SAN LEANDRO, SNELLINGS, MARIPOSA, BENTON, SANTA CLARA, REDWOOD CITY, OWENSVILLE, MILLERTON, SANTA CRUZ, GILROY, SCOTTSBURGH, MONTEREY, VISALIA, SAN ANTONIO, WOODVILLE, SAN LUIS OBISPO, HAVILAH,

KEYSVILLE, SANTA BARBARA, San BUENAVENTURA, LOS ANGELES, SAN BERNARDINO, WILMINGTON, SAN DIEGO, OAKLAND, ALAMEDA, BROOKLYN, DUTCH FLAT, FOLSOM, FORT YUMA, GRASS VALLEY, SAN FERNANDO, HEALDSBURG, PESCADERO, SONOMA, WATSONVILLE, ANTIOCH.

NEVADA.

VIRGINIA CITY, CARSON CITY, AUSTIN, DAYTON, UNIONVILLE,

WASHOE CITY, STILLWATER, LA PLATA, AURORA, BELMONT,

HUMBOLDT, GENEVA, LANDER CITY, CALLVILLE, ST. THOMAS,

MILL CITY, HIKO, HAMILTON, TREASURE CITY, ELKO.

MOUNTAIN RANGES.

Where situated? In what direction do they extend?

Sierra Nevada, Santa Cruz, Santa Lucia, Gavilan,
Coast Range, Siskiyou, San Rafael, Granite,
Pitt River, Toyabe, White, Humboldt?

MOUNTAIN PEAKS.

Where situated?

Sierra Nevada.

Mt. Whitney, Mt. Tyndall, Mt. Brewer, Mt. Lyell,
Mt. Shasta, Mt. Kaweah, Mt. Dana, Castle Peak?

Coast Range.

S. Bernardino, Pierce, Hamilton, Diablo,
Balley, Tamalpais?

RIVERS.

Where do they rise? What courses do they take? Into what waters do they flow?

Sacramento? Carson? Russian? Kern?
Klamath? Feather? American? Tulare?
Humboldt? Trinity? Tuolumne? Mohave?
San Joaquin? Eel? Mariposa? Salinas?
Fresno? Merced? Stanislaus? Santa Clara?
Reese? Truckee? Walker? Napa?

LAKES.

Where are they? What are their inlets and outlets?

Tulare? Goose? Pyramid? Carson?
Lower Klamath? Owens? Walker? Franklin?
Mono? Mud? Winnemucca? Kern?
Tahoe? Clear? Wright? Eagle?
Buena Vista? Honey? Mohave? Rhett?

CAPES OR POINTS.

From what counties do they project?

Mendocino? Arena? Loma? Concepcion?
St. George? Reyes? San Luis? Vincent?
Gold Bluff? Point Pinos? Pt. Arguilla? Point Sal?

BAYS.

Where are they? Into what waters do they open?

Trinidad? San Francisco? Estero? San Pedro?
Humboldt? Half Moon? San Luis? San Diego?

STRAITS AND CHANNELS.

What lands do they separate? What waters do they connect?

Golden Gate? Santa Barbara Channel?

ISLANDS.

Where are they? By what waters are they surrounded?

Farallone? San Miguel? San Clemente? Santa Barbara?
Santa Cruz? Santa Rosa? San Nicolas? Santa Catalina?

DESCRIPTIVE GEOGRAPHY.

1. *CALIFORNIA* is situated in the western part of the United States, on the Pacific Coast.

2. *It Extends* from Oregon on the north to Lower California on the south.

3. *Its Length* is about 750 miles, its average breadth 250 miles, and its area about 189,000 square miles.

4. *In Size*, it is the second State in the Union, Texas being the largest. It is about as large as the Eastern and Middle States combined.

5. *Its Northern Boundary* is the parallel of 42° north latitude, and is nearly *in a line* with the northern boundaries of Indiana, Ohio, Pennsylvania, Connecticut, and Rhode Island.

6. *The State lies in the same General Direction* as its coast line and mountain ranges.

7. *The principal Mountain Ranges* are two, the Sierra Nevada in the eastern, and the Coast Range in the western part of the State; these ranges unite in the northern and southern parts of the State.

8. *These two Ranges enclose* the great valley of California, which is drained by the two largest rivers in the State, the Sacramento and San Joaquin.

9. *The Coast Mountains* are near the coast, and rise to heights varying from 2,000 to 4,000 feet. In the northern part of the State they are covered with luxuriant forests.

10. *Between their Ridges are* numerous valleys noted for their beauty, richness, and salubrity (see page 37, paragraphs 23 and 38).

11. *The Sierra Nevada* (snowy range), which extends along the eastern part of the State, rises generally above the snow limit, having many peaks varying from 7,000 to 15,000 feet in height.

12. *The most Noted Peaks in the State* are, in the Sierra Nevadas, Mt. Whitney, 15,086 feet high; Mt. Shasta, 14,442 feet; Mt. Tyndall, 14,386 feet; Mt. Kaweah, 14,000 feet; Mt. Dana, 13,227 feet; Mt. Lyell, 13,217 feet; and, in the Coast Range, Mt. San Bernardino, 8,370 feet; Mt. Balley, 6,357 feet; Mt. Pierce, 6,000 feet; Mt. Hamilton, 4,450 feet; and Mt. Diablo, 3,876 feet.

13. *The Principal Valleys of the State* are the Sacramento, San Joaquin, Santa Clara, Pajaro, Salinas, Shasta, Scott, Napa, Amador, and Russian River.

14. *The Yo Semite Valley*, in the Sierra Nevada Mountains, is celebrated for the grandeur of its scenery. It is formed by the Merced River, and is situated in Mariposa County, 250 miles from San Francisco. It is about 8 miles long, and from half a mile to one mile wide, enclosed by precipitous walls of rock, rising at one point 4,737 feet above the level of the river.

The highest water-fall known in the world is the Yo Semite, on the north side of this valley, descending in three falls, 2,600 feet, the highest being 1,500 feet.

15. *The Geysers*, in Sonoma County, are hot springs, which emit water and steam containing various salts.

16. *Lakes.* Tulare Lake is the largest lake in the State. Lake Tahoe is 6,000 feet above the level of the sea, and is remarkable for the clearness of its water. The water of Mono Lake is exceedingly salt. Borax Lake is so named from the presence of borax in large quantities in the mud at its bottom.

17. *Bays.* San Francisco Bay, communicating with the Pacific Ocean by the Golden Gate, is about 60 miles long and 10 miles wide. It affords an extensive and excellent harbor. San Pablo and Suisun Bays are properly continuations of San Francisco Bay, the latter receiving the Sacramento and San Joaquin Rivers. Humboldt Bay, about 12 miles long and 4 miles wide, is a great lumber shipping port. Wilmington Bay, at the head of San Pedro Bay, is the shipping point for Los Angeles and San Bernardino Counties. San Diego Bay is, next to San Francisco Bay, the best harbor in the State.

18. *Islands.* The Farallone Islands, 23 miles outside the Golden Gate, belong to San Francisco County. Santa Cruz, San Miguel, Santa Rosa, and San Nicolas are chiefly valuable for sheep raising. Most of the Bay Islands belong to the United States, and are used for harbor defence. The United States Navy Yard is situated on Mare Island.

19. *The Climate* of California is milder and more equable than that of the Atlantic or Central States in the same latitude. The mean temperature of San Francisco in September is 58 degrees, and in January 50 degrees, being a difference of only 8 degrees between the warmest and coldest months.

20. *The Summers* are dry, and the winters rainy (see page 37, paragraphs 21, 22, 23, 31 and 38).

21. *In the South-eastern Part* of the State is a desert region where the heat is intense.

22. *The Agricultural Productions* are varied and abundant, including those of both temperate and tropical regions. Aside from the great mineral wealth of the State, its soil and climate render it one of the richest countries in the world.

23. *The Principal Agricultural Productions* are wheat, barley, grapes, sugar beet, hops and various kinds of fruits. The yield of wheat is about 30,000,000 bushels annually: of barley over 7,000,000 bushels; and of wine, about 5,000,000 gallons. The rearing of silk-worms receives considerable attention.

24. *The Forests* furnish valuable timber, including redwood, oak, pine, laurel, and cedar.

25. *The Big Trees* (*sequoia gigantea*), a species of redwood, are found in several groves, the most noted being in Calaveras County. The largest trees are about 30 feet in diameter and about 350 feet in height.

26. *Sheep Raising* is an important interest in California and on some of the neighboring islands. In 1874 the product was more than 36,000,000 pounds of wool.

27. *The leading Mineral Productions* are gold, mercury, silver, and copper; besides these are iron, platinum, coal, nickel, salt, borax, lead, tin, zinc, etc.

28. *The Methods of Mining Gold* are three: quartz mining, placer mining, and hydraulic mining.

29. *The Value of the Gold* taken from California since 1849 is $1,000,000,000. The largest amount in one year was in 1853, $65,000,000.

30. *The most Noted Quicksilver Mine* is at New Almaden, Santa Clara County, producing about 2,000,000 pounds a year.

31. *The Best Coal* in the State is found on Mount Diablo.

rests of the State have ad-
1870, to $66,000,000. They
ds, flour, sugar, iron, lumber,
uo, carriages—in short, nearly
tate.

nia is of great importance, and
been greatly promoted by the
ers to China and Japan, and
ilroad, thus facilitating trade
sia, by way of San Francisco.
y the construction of a canal
The State has also an increas-
slands, Australia, Mexico, and

of the State are gold, grain,

raveling throughout the State
nd rapidly increasing system
er steamers.

polis of the Pacific coast, is
part of a peninsula which is
San Francisco and the Pacific
1835, and was formerly called

kable for the rapidity of its
in 1845 numbered but 150, is

city is commerce. Its manu-
ensive.

Francisco are numerous and
the school-houses are among

s the second city in the State
east bank of the Sacramento
merican, in the midst of one of
the State. It is the western
Railroad, and has direct com-
both by water and rail. The
troyed by floods, but is now
capitol is a costly and elegant

slough near its junction with
iles from San Francisco by
is the distributing point for a
one of the leading grain ports

ly situated at the junction of
r. It has important railroad
and prosperous city.

ate capital, is in Santa Clara
f San Francisco Bay. It has
of the pleasantest cities in the

excellent harbors. Both have

roville, and Dutch Flat are
ns.

Alameda are pleasantly situ-
ated on the eastern side of San Francisco Bay, and are closely connected with San Francisco by ferries and rail.

47. *Los Angeles and San Diego* are the principal towns in the southern part of the State. The former is in the midst of an excellent fruit country. Grapes, oranges, lemons, and other tropical fruits abound. The latter is the oldest town in the State, having been founded in 1759. It is growing rapidly, and is the proposed terminus of the Southern Pacific Railroad.

48. *History.*—California was discovered in the sixteenth century. It formed a portion of Mexico until it was ceded to the United States in 1848, at the close of the Mexican war. Gold was discovered near Coloma in the same year, and since that time the growth of the State has been very rapid. It was admitted to the Union as the thirty-first State, in 1850.

49. *Government.*—The Governor and Senators hold office four years; the members of the Assembly two years; the Judge of the Supreme Court ten years; and of the County Courts four years. All other State officers hold office four years.

50. *Education.*—Public Instruction is under the charge of one State Superintendent, elected by the people every four years, and one County Superintendent in each County, who holds office two years. The educational system of the State is complete, ranging through all grades, from Primary to the State University. There are also many private and denominational institutions of learning in the State.

51. *NEVADA extends* from Oregon and Idaho on the north, to the Colorado River on the south, a distance of 500 miles, and from California on the west to Utah on the east, a breadth of 300 miles. Its area is about 112,000 miles.

52. *Surface.*—The State is principally a vast basin, diversified by mountains, valleys, and plateaus.

53. *The Great Basin* is partly in this State: its elevation is from 4,000 to 5,000 feet above the sea level.

54. *The Mountain Ranges* are short and numerous, generally extending nearly north and south. The East Humboldt Range is the highest. There are several peaks in the State ranging from 8,000 to 12,000 feet in height, but few of them have been measured.

55. *The Rivers* are small, and empty into lakes or "sinks," which have no connection with the ocean; the largest river is the Humboldt. The lakes, having no outlets, are generally salt or alkaline.

56. *The Climate* is generally dry and the soil barren; agriculture is mostly carried on by means of irrigation. The north-western part of the State is a desert.

57. *Its Chief Source of Wealth* is in its silver mines, which are found in various parts of the State, particularly in Washoe and Storey counties in the west, and the White Pine region in the centre, the latter comprising a district about 12 miles square, in the White Pine Mountains, where recent discoveries of silver have attracted much attention. Gold, silver, copper, lead, iron, and salt are also found.

58. *The Leading Towns* are Virginia City, Carson City, Gold Hill, Elko, Belmont, Austin, Treasure City, and Hamilton. The last two are in the White Pine Silver Mining District, Treasure City being about 9,000 feet above the sea level.

59. *Nevada was admitted* as a State in 1864.

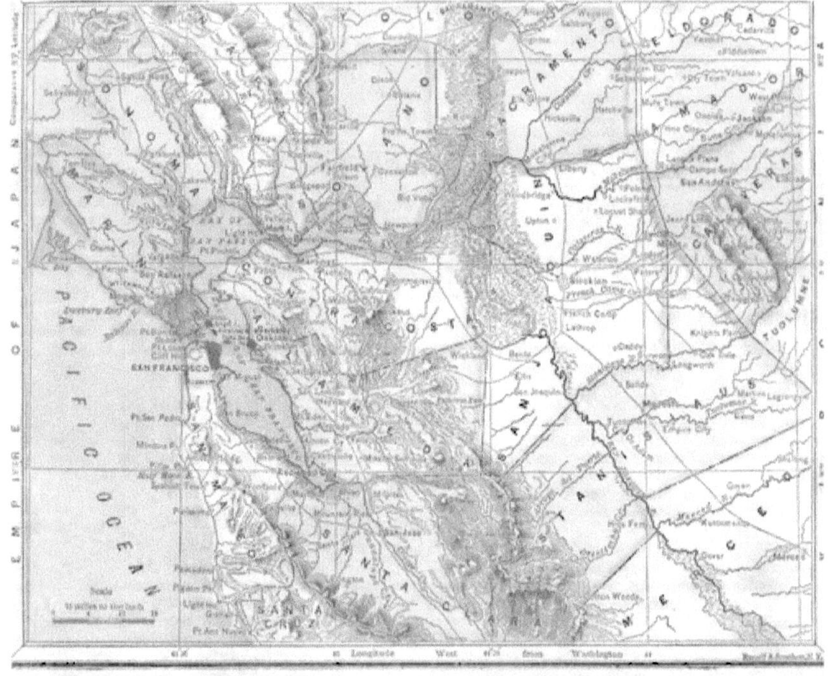

EXERCISES ON THE MAP.

What is the largest city west of the Rocky Mountains?

Through what gate or strait would you sail in going from San Francisco to the ocean?

What is the length of San Francisco Bay? Its greatest width?

What bay north of San Francisco Bay?

Give the length and breadth of the Bay of San Pablo.

What bay east of the Bay of San Pablo?

What two large rivers empty into Suisun Bay? Describe them.

In what county is the city of San Francisco?

What county south of San Francisco county?

What county south of San Mateo? South-east?

What county borders on the easterly side of the Bay of San Francisco? On the southerly side of Suisun Bay? On the northerly side?

What county between the Bay of San Pablo and the Pacific?

What county north of Marin county? What county east of Sono'ma? What south and east of Napa county? North and north-east of Solano?

Through what counties does the San Joaquin River flow? Name and describe its tributaries.

In what direction does the land east of the San Joaquin slope?

In what direction does the land slope which lies between the San Joaquin River and the Coast Range?

Where is the highest land of Santa Clara county? The lowest?

What mountains on its eastern side? Western? What high peaks has it?

What rivers flow through San Joaquin county? Stan'islaus county? Merced county? Santa Clara county? San Mateo county? Sonoma county? Napa county? Santa Clara county?

What is the county town of the following counties: Marin? Sonoma? Napa? Solano? Yolo? Sacramento? Amador? Calaveras? San Joaquin? Contra Costa? Alameda? Santa Clara? San Mateo? Stan'islaus?

What high peak in Contra Costa? In Marin county?

Where is Mare Island? Alcatraz? Angel? Goat Island?

Where is Hunter's Point? Point Pinole? Saucelito Point? Rincon Point? Point Bonita? Lime Point? Point Lobos?

MOUNTAINS.

Where are they? In what direction do the ranges extend?

Coast Range? Santa Cruz? Matacamas? Mt. Diablo?
Mt. Hamilton? Black Mt.? Mt. Lewis? Mt. Tamalpais?

RIVERS AND CREEKS.

Where do they rise? In what directions do they flow, and into what waters?

Calave'ras R.? (ras) Mokelumne? Napa? Tuolumne?
Calaveras Cr.? Cosumnes? Coyote Cr.? Dry Cr.?
Stanislaus? Merced? Pescadero? Guadalupe?

CITIES AND TOWNS.

In what part of what county? On or near what water?

SAN FRANCISCO, NEW YORK, PESCADERO, PACHECO,
SACRAMENTO, LAKEVILLE, PETALUMA, JACKSON,
STOCKTON, SAN LEANDRO, HAYWARDS, IONE CITY,
SAN JOSE, REDWOOD CITY, VALLEJO, RIO VISTA,
SAN RAFAEL, FAIRFIELD, BENICIA, SUISUN,
MARTINEZ, SAN ANDREAS, ANTIOCH, SUTTERVILLE,
NAPA, KNIGHT'S FERRY, MOKELUMNE CITY, SAN LORENZO,
TUOLUMNE CITY, OAKLAND, LIBERTY, SANTA CLARA,
SAN MATEO, NEW ALMADEN', MENLO PARK, ALAMEDA,
COPPEROPOLIS, SANTA ROSA, MAYFIELD, BROOKLYN.

DESCRIPTIVE GEOGRAPHY.

1. ***OREGON is situated*** on the Pacific Coast, in the northwestern part of the United States.

2. ***It lies in a line*** directly west of New England, and its southern boundary is the parallel of 42 degrees north latitude, which parallel is the dividing line between New York and Pennsylvania.

3. ***The Length of Oregon*** from east to west is about 350 miles, its breadth 300 miles, and its area is about 100,000 square miles, being equal to that of New York and Pennsylvania combined.

4. ***The Mountain Ranges*** are three: the Coast Range, the Cascade Range, and the Blue Mountains. They extend north and south across the State.

5. ***The Coast Range*** extends along the coast, *and is pierced* by numerous streams, which empty into the Pacific. Their height varies from 2,000 to 4,000 feet.

6. ***The Cascade Mountains*** are about 120 miles from the coast, and extend through Oregon and Washington. They also extend into California, where they are called the Sierra Nevada. Their height varies from 4,000 to 13,000 feet.

7. ***The Principal Peaks*** of this range are, Mount Hood, an extinct volcano, Mount Jefferson, Mount Pitt, and the Three Sisters; all of which rise above the limit of perpetual snow.

8. ***The Blue Mountains*** are in the eastern part of the State, and have short ridges extending east and west at right angles to the main ridge.

9. ***The State is divided*** into three physical sections by the Blue and the Cascade Mountains,—the Western, Middle, and Eastern, styled, respectively, the Lower, Middle, and Upper Countries.

10. ***The Western Section*** is between the Pacific Ocean and the Cascade Mountains, and covers about one-third the area of the State. Its beautiful and fertile valleys contain nearly the whole of the tillable land and all the principal cities and towns in the State.

11. ***The Middle Country*** is an elevated plain, useful in some places for pasture; but its southern portion is salt and barren.

12. ***The Upper Country lies*** east of the Blue Mountains, and is generally dry and barren; rich, however, in mineral wealth.

13. ***The Principal Harbors*** are those afforded by the Columbia and Umpqua Rivers, and the Tillamook, Yaquin'a, and Coos Bays.

14. ***The Principal River*** is the Columbia, the largest river in America which empties into the Pacific Ocean. In its course from British America to the Cascades, the head of navigation, rapids and waterfalls are numerous. It supplies salmon in abundance. Its most important branch is the Willamette. The Umpqua and Rogue Rivers afford an outlet to a valuable lumber region.

15. ***The Willam'ette Valley*** is the largest and most fertile in the State. It contains 2,000,000 acres of excellent farming land. Its length from north to south is about 120 miles, and its average width 50 miles, being one-third larger than the State of Connecticut. The Umpqua and Rogue River Valleys are also important.

16. ***The Climate*** of the Western Section, or Lower Country, is much milder and more uniform than that of corresponding latitudes on the Atlantic coast, owing to the warm, moist winds which blow from the Pacific.

Rain is abundant because of the cooling influence upon those winds of the Cascade Mountains. (See page 34, paragraph 41, and page 37, paragraphs 31–38.) The greater portion usually falls during the months of November, December, March, and April. Rain seldom or never falls in harvest time—from the first of August to the middle of September.

17. ***The Climate of the Middle and Upper Countries*** is dry, and subject to great extremes of heat and cold.

18. ***The Principal Agricultural Productions*** are wheat, barley, oats, garden vegetables, and orchard fruits. The trade in wool and live stock is important.

19. ***The Forests*** of the Lower Country abound in magnificent pines, with fir, oak, hemlock, cedar, maple, and other trees valuable for timber.

20. ***The Mineral Productions*** consist of gold, silver, copper, iron, and coal.

21. ***Salt Springs*** are numerous.

22. ***Manufactures*** receive considerable attention, and comprise lumber, flour, and woolen goods.

23. ***The Export Trade*** with California and the Eastern States is important. Grain is shipped to England, and regular lines of transportation are established with New York

24. ***The Principal Exports*** are grain, flour, wool, lumber, apples, dried fruits, hides, and pickled salmon.

25. ***Portland***, the principal city, is beautifully located on the west bank of the Willamette River, 15 miles from its mouth, at the head of ship navigation. It is rapidly increasing in wealth and population, and is the second city in importance on the Pacific slope.

26. ***Salem***, the capital, is finely situated in a rich prairie, on the east bank of the Willam'ette.

27. ***Oregon City*** is on the right or east bank of the Willam'ette River, about 10 miles south of Portland. The falls in the river furnish immense water-power for manufacturing purposes.

28. ***The other Important Towns*** are Albany, Corvallis, Eugene City, Dalles, Astoria, and Jacksonville.

29. ***Public Education*** has received much attention. Besides the excellent public schools, there are colleges at Salem, Forest Grove, and Corvallis, and numerous academies throughout the State.

30. ***The Legislature*** assembles every two years. The members of the Senate are elected for four years, and of the House for two years.

31. ***Oregon was Organized as a Territory*** of the United States in 1848, when it extended northward to British America, and eastward to the Rocky Mountains.

32. ***From its Northern half***, Washington Territory was organized in 1853; and in 1859, Oregon was admitted as a State with its present boundaries.

EXERCISES ON THE MAP.

OREGON,—by what is it bounded on the north? On the east? On the south? On the west?

What is its length? Its breadth? *(Apply the scale of miles.)*

What mountain ranges in Oregon?

Is the greater part of the State east or west of the Cascade Range?

What prominent peaks in the Cascade Range?

What rivers east of the Blue Mountains?

In what direction does the surface of that part of the State slope?

What rivers between the Cascade Range and the Blue Mountains? Into what do those rivers empty?

How does the surface of the middle section of the State slope?

What rivers between the Cascade Range and the Pacific? Into what do they flow? Which is the largest?

In what direction does the land slope which is drained by the Willamette? By the Umpqua and Rogue Rivers?

Mention the largest lakes in Oregon? Which have outlets? Which have no outlets? In what part of the State are lakes the most numerous?

What counties border on the Columbia? On the Snake River?

What counties border on the Pacific? On California? On Nevada?

What counties between the Cascade Range and Willam'ette River?

What counties border on the west side of the Willam'ette?

Through what counties does that river flow?

Name all the counties in the Willamette Valley? In the Umpqua Valley? In the Rogue River Valley?

By what is Coos County drained?

What county towns on the south bank of the Columbia? On the east bank of the Willamette? On its west or left bank? Between the Willamette and the Coast Range? In the Valley of the Umpqua? In the Valley of Rogue River?

What towns on the Pacific Coast?

Draw a map of Oregon, as directed on page 105.

WASHINGTON,—by what is it bounded on the north? East? South? West?

What is its length? Its breadth?

What mountains extend through the territory?

Mention the principal peaks? The largest river?

What are the two principal forks or tributaries of the Columbia River?

Where are their sources?

What tributaries has the Columbia from the west?

In what direction does the land slope which lies between the Columbia River and the Cascade Range? Between the Columbia River and the Rocky Mountains?

What rivers west of the Cascade Mountains?

Into what do they flow?

What counties border on British America? On the Pacific? On the Strait of Ju'an de Fu'ca and inlets?

What counties in the south border on the Columbia River? What county in the southeast corner of the territory? At the mouth of the Columbia?

What towns on the north bank of the Columbia?

What towns in the northwest?

Draw a map of Washington, as directed on page 105.

IDAHO,—by what is it bounded on the north? East? South? West?

What ranges of mountains on the east?

By what river and its tributaries is nearly the whole of Idaho drained?

By what rivers is the northern part drained?

On which side is the highest land? The lowest?

In what direction does the land south of the Snake River slope?

(Observe the course of the rivers.)

What rivers flow through that portion of the State?

What rivers flow westerly into the Snake River?

What branch of the Colorado river has its sources near those of the Snake River?

What lakes in the northern part? In the eastern part?

What is the extent of Idaho from north to south? From east to west?

What county forms the northern part of Idaho?

What county borders on Oregon and Nevada? On Utah?

What counties are bounded partly by the mountains? By the Snake River?

What county in the centre of the State?

Draw a map of Idaho, as directed on page 105.

MONTANA,—by what is it bounded on the north? South? West?

What high mountains extend through its western part?

What mountains on its western border?

What great river has its sources in the southwestern part of Montana?

On which side of the Rocky Mountains are the sources of the Missouri?

What small rivers form its headwaters?

What large rivers on the Pacific slope have their sources near those of the Missouri?

Mention the principal passes over the Rocky Mountains in Montana?

What rivers flow into the western side of the Missouri? Into the eastern side?

What lake in the northwest?

What towns in Madison and Beaver Head Counties?

What towns between the Missouri River and the Rocky Mountains?

What towns west of the Rocky Mountains?

Draw a map of Montana, as directed on page 105.

REVIEW

MOUNTAINS.

Where are they? In what direction do the ranges extend?

Cascade Range? Cedar Mt.? Mt. Pitt?
Rogue River Mts.? Mary's Peak? Mt. Adams?
Three Sisters? Mt. St. Helen's? Rocky Mts.?
Mt. Baker? Three Buttes? Mt. Hood?
Mt. Olympus? Blue Mts.? Scott's Peak?
Coast Range? Mt. Jefferson? Mt. Rainier?

RIVERS.

Where do they rise? In what directions do they flow, and into what waters?

Columbia? Ump'qua? Fall? McKenzie's?
Snake? Rogue? John Day's? Malheur?
Willam'ette? Owy'hee? Powder? Grande Ronde?
Chehalis? Skagit? U-ma-til'la? Salmon?

LAKES.

Where are they? What are their inlets and outlets?

Upper Klamath? Harney? Malheur? Flathead?

CAPES.

From what counties do they project?

Lookout? Foulweather? Arago? Blanco?
Flattery? Disappointment? Pt. Adams?

BAYS.

Where are they? Into what waters do they project?

Tillamook? Gray's Harbor? Yaquina? Coos?
Puget Sound? Shoalwater?

CITIES AND TOWNS.

In what part of what county? On or near what water?

PORTLAND. OREGON CITY. CANON CITY. EUGENE CITY.
SALEM. ASTORIA. OAKLAND. HARRISBURG.
ALBANY. EMPIRE CITY. ROSEBURG. AURORA.
AUBURN. ST. HELENS. DALLES. SILVERTON.
CORVALLIS. CANONVILLE. DALLAS. UMATILLA.
LA GRANDE. UNION. GRANITE CITY. ELLENSBURG.
JACKSONVILLE. PORT ORFORD. UMPQUA CITY. FOREST GROVE.
OLYMPIA. PORT TOWNSEND. VIRGINIA CITY. IDAHO CITY.
STEILACOOM. VANCOUVER. BANNOCK CITY. LEWISTON.
SEATTLE. WALLA WALLA. GALLATIN. SILVER CITY.
WHATCOM. HELENA. BOISE CITY. MONTICELLO.

DESCRIPTIVE GEOGRAPHY.

1. ***WASHINGTON TERRITORY*** occupies the most north-western portion of the United States, except Alaska; its northern boundary being in a line with that of Minnesota.

2. ***It lies*** between the parallels of 46° and 49° north latitude, between which are, also, Lake Superior, Northern Maine, Central France, Switzerland, and Austria.

3. ***The length*** of Washington is about 350 miles from east to west, and its breadth from north to south is about 225 miles. Its area is about 70,000 square miles.

4. ***The Principal Mountain Ranges*** are the Cascade Range and the Coast Range, extending nearly north and south. The principal peaks are Mt. Rainier, Mt. Baker, Mt. Adams, and Mt. St. Helens, all of which rise above the snow limit.

5. ***The Cascade Range*** in Washington and Oregon is a continuation of the Sierra Nevada, of California. Their height is about 6,000 feet, and their distance from the coast about 100 miles. This range divides Washington Territory into two physical regions—the **western or lower, and the eastern or higher.**

6. ***The Western section*** contains rich and well-watered valleys, vast evergreen forests of fir, spruce, cedar, and hemlock, and nearly all the cities and towns in the Territory.

7. ***The Eastern section*** is chiefly dry and stony, although it contains some excellent pasture grounds. Walla Walla Valley, in the south-east, is well adapted to agriculture and grazing.

8. ***The Climate*** of Washington Territory is similar to that of the corresponding sections of Oregon, being much milder than in the same latitudes on the Atlantic slope.

9. ***The Principal Valleys*** are the Puget Sound Valley, or basin, the Columbia Valley, and the Walla Walla Valley. The former contains over 10,000 square miles.

10. ***Puget Sound affords*** the chief commercial outlet for the Territory, having about 1,500 miles of water front, with many excellent harbors. It is surrounded by valuable timber, great quantities of which are exported annually. Coal is supplied from Bellingham Bay, and oysters are found in abundance in Shoalwater Bay.

11. ***The Chief Exports*** are lumber, ship timber, coal, and oysters.

12. ***Gold*** is found in the eastern part of the Territory.

13. ***The Principal Towns*** are Olympia, the capital, Steilacoom, Seattle, Port Townsend, Monticello, Vancouver, and Walla Walla.

14. ***Washington was organized*** as a Territory in 1853.

15. ***IDAHO*** is somewhat triangular in shape, about 480 miles in extent from north to south, and 300 miles wide on its southern border. Its area is about 95,000 square miles.

16. ***The Rocky Mountains*** and a branch, or continuation, called the Bitter Root Mountains, are on its eastern border. Among other mountains in the Territory, are the Salmon River and Bear Mountains, the Three Buttes and the Three Tetons.

17. ***The Principal Rivers*** in Idaho are the Snake or Lewis, and its tributaries the Salmon and Clearwater, the general slope of the surface being toward the west.

18. ***Its Principal Productions*** are gold and silver.

19. ***The Soil*** of large portions is well adapted to farming and grazing, and water-power for manufacturing purposes is abundant.

20. ***The Chief Towns*** are Boise City, the capital, Lewiston, Idaho City, Ruby City, Oro Fino, and Silver City.

21. ***Idaho was organized*** as a Territory in 1863.

22. ***MONTANA*** is in the north-western part of the United States, lying between Dakota on the east and Idaho on the west; on the north it adjoins British America, from which it is separated by the parallel of 49°, here the northern boundary of the Union.

23. ***It lies*** partly on the Atlantic and partly on the Pacific slope; the Rocky Mountains, the great water-shed, extending through its western part.

24. ***The Pacific Slope*** includes the north-western portion, which is drained by the Clarke's River and its tributaries.

25. ***The Largest Portion of Montana*** is embraced in the Atlantic slope, and is drained by the Missouri River, which has its headwaters in the south-western part of the Territory.

26. ***The Greatest Length*** of Montana from east to west is 500 miles, and its breadth, from north to south, about 270 miles.

27. ***Montana*** is rich in gold and silver, which are extensively mined in the west and south-west, where, consequently, are the most thickly settled portions of the Territory. It also contains copper, iron, and coal of superior qualities.

28. ***The Climate*** is pure, healthful, and delightful, and much dryer than that of other parts of the United States.

29. ***The Soil*** of its extensive valleys is very productive, and its agricultural and grazing facilities are excellent.

30. ***Its Trade*** with the States eastward, by way of the Missouri River, is considerable and constantly increasing.

31. ***Its Population*** is about 40,000.

32. ***The Great Falls*** of the Missouri, remarkable for their grand and picturesque scenery, are about 30 miles above, or south-west of, Fort Benton, the nominal head of steam navigation. They comprise a succession of cataracts and rapids, with which the river is filled for about 12 miles, having an aggregate descent of 400 feet. The principal cataract is 87 feet high and 900 feet wide.

33. ***Helena***, the metropolis and commercial centre, is 140 miles south-west of Fort Benton. It is situated on a creek which flows into the western side of the Missouri River.

34. ***Virginia City***, is situated in the south-western part of the Territory, about 130 miles south of Helena.

35. ***Montana was Organized*** as a Territory in 1864.

TERRITORIES.	CAPITALS.	TERRITORIES.	CAPITALS.
WASHINGTON ...	Olympia.	UTAH	Salt Lake City
IDAHO	Boise City.	ALASKA..........	Sitka.
MONTANA........	Helena.	ARIZONA.........	Tucson.
DAKOTA..........	Yankton.	NEW MEXICO....	Santa Fé.
WYOMING........	Cheyenne.	INDIAN	Tahlequah.

COMPARATIVE SIZES—KANSAS, THE COMMON MEASURE AND FRAME.

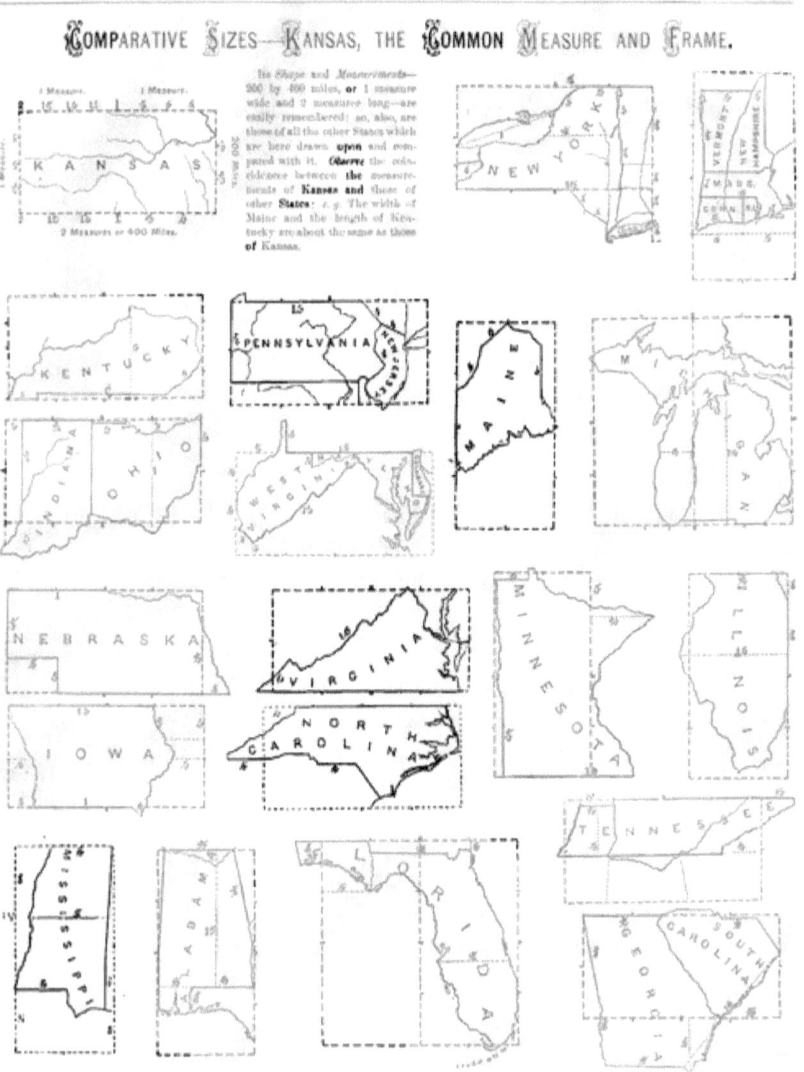

Its *Shape* and *Measurements*—200 by 400 miles, **or** 1 measure wide and 2 measures long—are easily remembered; so, also, are those of all the other States which are here drawn **upon** and compared with it. ***Observe*** the coincidence between **the** measurements of **Kansas and** those of other **States**; *e. g.* The width of Maine and the length of Kentucky are about the same as those **of** Kansas.

MONTEITH'S RELIEF MAP of the UNITED STAT

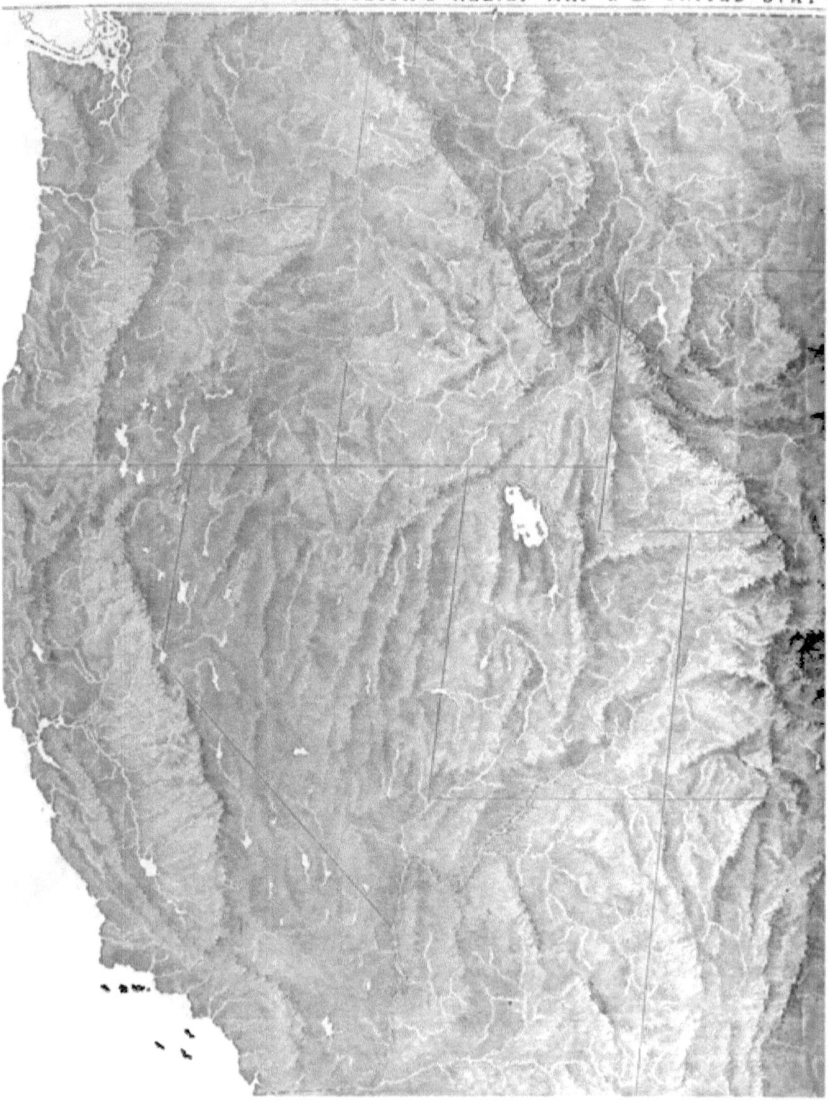

from the PACIFIC OCEAN to the MISSISSIPPI RIVER.

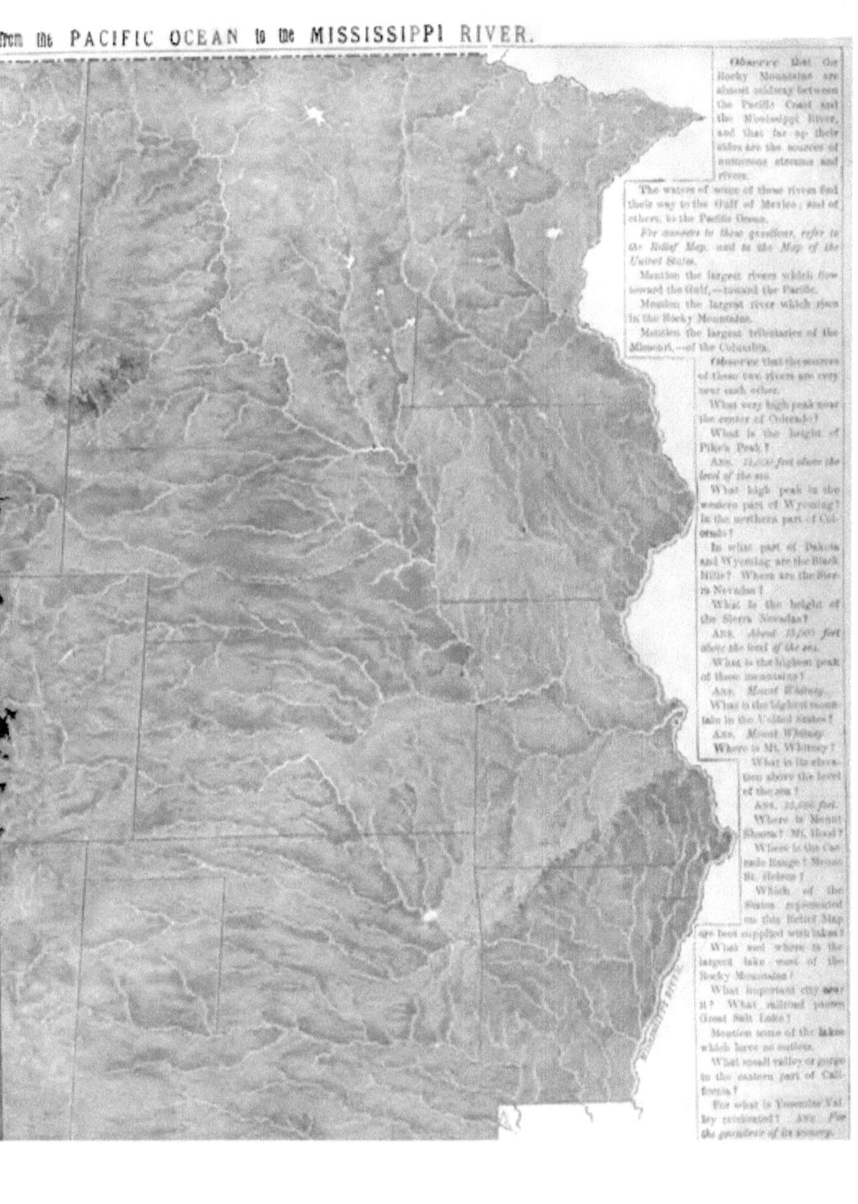

Observe that the Rocky Mountains are almost midway between the Pacific Coast and the Mississippi River, and that far up their sides are the sources of numerous streams and rivers.

The waters of some of these rivers find their way to the Gulf of Mexico; and of others, to the Pacific Ocean.

For answers to these questions, refer to the Relief Map, and to the Map of the United States.

Mention the largest rivers which flow toward the Gulf,—toward the Pacific.

Mention the largest river which rises in the Rocky Mountains.

Mention the largest tributaries of the Missouri,—of the Columbia.

Observe that the sources of these two rivers are very near each other.

What very high peak near the center of Colorado?

What is the height of Pike's Peak?

Ans. *14,000 feet above the level of the sea.*

What high peak in the western part of Wyoming? In the northern part of Colorado?

In what part of Dakota and Wyoming are the Black Hills? Where are the Sierra Nevadas?

What is the height of the Sierra Nevadas?

Ans. *About 15,000 feet above the level of the sea.*

What is the highest peak of these mountains?

Ans. *Mount Whitney.*

What is the highest mountain in the United States?

Ans. *Mount Whitney.*

Where is Mt. Whitney?

What is its elevation above the level of the sea?

Ans. *15,000 feet.*

Where is Mount Shasta? Mt. Hood?

Where is the Cascade Range? Mount St. Helens?

Which of the States represented on this Relief Map are best supplied with lakes?

What and where is the largest lake west of the Rocky Mountains?

What important city near it? What railroad passes Great Salt Lake?

Mention some of the lakes which have no outlets.

What small valley or gorge in the eastern part of California?

For what is Yosemite Valley celebrated? Ans. *For the grandeur of its scenery.*

HEIGHTS OF MOUNTAINS.

ASIA.

Mountains.	Situation.	Elevation in feet.
Mt. Everest	In the Himalaya Chain, the highest point on the globe	29,000
Mt. Kinchinjunga	In the Himalaya Chain	28,178
Mt. Dhawalagiri	" " "	28,000
Mt. Chumalari	" " "	23,929
Mt. Hindoo Kosh	Hindoo Coosh Mts.	20,230
Thian-Shan Mts.	Chinese Empire	20,000
Kuen-Lun Mts.	North of Thibet ... 16,000 to	18,000
Mt. Ararat		17,260
Mt. Klutchevskaja	Kamtchatka	15,763
Mt. Demavend	Elburz Mts., south of the Caspian Sea	14,700
Mt. Argish	Anti-Taurus Mts., Asia Minor	13,000
Soliman Mts.	between Hindostan and Afghanistan	11,000
Lebanon Mts.	Syria	11,000
Altai Mts.	Highest	10,700
Yablonoi Mts.	Highest	10,000
Mt. Hermon	highest of Anti-Lebanon Mts., Syria	10,000
Mt. Sinai	Arabia	7,497

SOUTH AMERICA.

Mountains.	Situation.	Elevation in feet.
Vol. Aconcagua	west of Chili	23,910
Mt. Sahama	south part of Peru	22,350
Vol. Gualatieri	" " "	22,000
Mt. Chimborazo	Ecuador	21,424
Mt. Sorata	Bolivia	21,286
Mt. Illimani	"	21,149
Mt. Chequibamba	Peru	21,000
Vol. Arequipa	"	20,320
Vol. Atacama	Bolivia	19,000
Vol. Antisana	Ecuador	19,150
Vol. Cotopaxi	"	18,877
Vol. Tolima	Colombia	18,000
Vol. Pichincha	Ecuador	15,924
Andes of Chili	highest	15,000
Andes of Patagonia	"	6,000
Cape Horn		500
Height of Snow Line	in Colombia	15,800
" " "	in Bolivia ... 16,000 to	18,000
" " "	in Chili	8,500
" " "	in Patagonia	3,000

NORTH AMERICA.

Mountains.	Situation.	Elevation in feet.
Vol. Popocatepetl	Mexico	18,200
Mt. St. Elias	Alaska ... about	18,000
Vol. Orizaba	Mexico	17,371
Mt. Brown	British America	15,900
Mt. Hooker	" "	15,625
Mt. Whitney	California	15,088
Mt. Fairweather	Alaska	14,900
Pike's Peak	Colorado	14,500
Mt. Shasta	California	14,442
Mt. Tyndall	"	14,386
Mt. Kaweah	"	14,000
Fremont's Peak	Wyoming Territory	13,570
Mt. St. Helens	Washington Territory	13,400
Mt. Dana	California	13,227
Mt. Lyell	"	13,217
Long's Peak	Colorado	12,500
Mt. Hood	Oregon	11,225
Mt. Lola	California	10,000
Mt. San Bernardino	"	8,500
Mt. Baker	Washington Territory	
Mt. Rainier	" "	
Mt. Adams	" "	
Laramie Peak	Wyoming Territory	8,000
Blue Mts.	Jamaica	7,500
Sierra del Cobre	Cuba—highest	7,200
Mitchell's Peak	North Carolina	6,707
Mt. Washington	New Hampshire	6,428
Oraefa Jokull	Iceland	6,400
Mt. Bailey	California	6,357
Mt. Pierce	"	6,000
Vol. Hecla	Iceland	5,700
Vol. Soufrière	Guadaloupe	5,500
Mt. Marcy	New York	5,379
Mt. Katahdin	Maine	5,000
Mt. Hamilton	California	4,450
Mt. Mansfield	Vermont	4,430
Peaks of Otter	Virginia	4,260
Camel's Hump	Vermont	4,188
Saddleback Mt.	Maine	4,000
Mt. Diablo	California	3,856
Round Top	New York	3,804
Alleghany Mts.	average	2,500
Wachusett Mt.	Massachusetts	2,000
Highlands	New York	1,650
Pilot Knob	Missouri	1,500
Mt. Tom	Massachusetts	1,200

EUROPE.

Mountains.	Situation.	Elevation in feet.
Mt. Elbrooz	Highest of Caucasus	17,796
Mt. Blanc	" " Alps	15,810
Mt. Rosa	" in Switzerland	15,208
Mt. Cervin	between Switzerland and Italy	14,771
Mt. Pelvoux	France	14,100
Finster Aarhorn	Switzerland	14,026
Mt. Viso	between France and Italy	12,590
Ortler Spitz	Austria	12,812
Mt. Mulahacen	Spain	11,660
Mt. [illegible]		11,424
The Simplon	between Switzerland and Italy	11,541
Mt. Cenis	France	11,460
Maladetta	highest of Pyrenees	11,168
Peak of Valetta		11,300
Great St. Bernard	Switzerland	11,000
Cantabrian Mts.	Spain—highest	10,300
Mt. Perdu	Pyrenees	10,994
Vol. Etna	Sicily	10,874
Mt. Cenis	"	10,877
Mt. St. Gothard	Switzerland	10,300
Castile Mts.	Spain	10,552
Mt. Olympus	Turkey	9,745
Mt. Kora	"	9,375
Mt. Athos	Greece	9,026
Pic du Midi	Pyrenees	9,540
Pindus Mts.		8,950
Mt. Lomnitz	Austria	8,779
Monte Rotondo	Corsica	8,764
Mt. Guiona	Greece	8,239
Mt. Parnassus	"	8,068
Scandinavian Mts.	highest	8,150
Pass of Venasque	Pyrenees	7,917
Rhodope Mts.	Turkey	7,800
Sierra d'Estrella	Portugal	7,524
Mt. Gennargentu	Sardinia	7,000
Jura Mts.	France and Italy	6,588
Mt. Brie	France	6,200
Cevennes Mts.	"	5,794
Sierra Morena	Spain	5,500
Mt. Gloria		5,350
Fichtel Gebirge	Germany	5,000
Riesen Gebirge	"	5,260
Ural Mts.	highest	5,375
Vosges	France	4,688
Black Forest	Baden	4,675
Bohemian Mts.	Austria	4,613
Erz Gebirge	"	4,200
Ben Nevis	Scotland—highest in Great Britain	4,406
Ben Mac Dhui	"	4,327
Cairngorm	"	4,095
Ben Alder	"	4,000
Vol. Vesuvius	Italy	3,948
Ben Lomond	Scotland	3,915
Brocken Mt.	Harz Mts., Germany	3,740
Ben Wyvis	Scotland	3,700
Snowdon Mt.	Wales	3,571
Carn Tual	highest in Ireland	3,414
Sca Fell	England	3,092
Helvellyn	"	3,055
Skiddaw	"	3,022
Moravian Mts.	Austria	3,000
Height of Snow Line	Sierra Nevada, Spain	11,200
" " "	Caucasus Mts.	11,000
" " "	Sicily	9,500
" " "	Mt. Olympus	9,000
" " "	Swiss Alps	8,500
" " "	Central Italy	8,400
" " "	Pyrenees	8,200
" " "	Carpathians	6,000
" " "	Scandinavian Mts.	5,000
" " "	Grampians	4,500
" " "	North Cape	2,400

AFRICA.

Mountains.	Situation.	Elevation in feet.
Mt. Kenia	Eastern Africa	20,000
Mt. Kilimandjaro	" "	20,000
Abba Jared	Abyssinia	15,000
Cameroon Mts.	Lower Guinea	13,000
Peak of Teneriffe Vol.	Canary Islands	12,158
Mt. Miltsin	Morocco	11,400
Red Mts.	Madagascar	11,000
Mt. Spitzkop	Cape Colony	10,000
Table Mt.	Cape of Good Hope	3,570

OCEANICA.

Mountains.	Situation.	Elevation in feet.
Mt. Ophir	Sumatra	13,842
Mt. Kini Balu	Borneo	13,698
Vol. Mauna Loa	Sandwich Islands	13,760
Mt. Semeru	Java	12,000
Mt. Erebus	Victoria Land	12,400
Mt. Kosciusko	Australia	6,500
Mt. Humboldt	Tasmania	5,520

HEIGHTS OF SOME INHABITED SITES.

Names.	Situation.	Feet above sea level.
[illegible]	Peru	15,646
Tacora	Village in Peru	14,000
Antisana	Shepherds' huts, Ecuador	13,434
Potosi	City in Bolivia	13,330
Puno	City in Peru	12,870
[illegible]	Bolivia	11,900
Leh	City in Thibet	9,995
Quito	Ecuador	9,543
Sherman	highest point on Pacific R. R.	8,242
Hospice of Gt. St. Bernard	Alps	7,963
Arequipa	City in Peru	7,850
Mexico	City	7,400
Cabul	Afghanistan	6,396
Ispahan	Persia	4,140
Jerusalem	Palestine	2,600
Madrid	Spain	1,990
Munich	Bavaria	1,704
Geneva	Switzerland	1,230

DISTANCES AT WHICH MOUNTAINS HAVE BEEN SEEN.

Mountains.	Miles.
Himalaya Mts.	224
Mt. Ararat	240
Mt. Chimborazo	150
Peak of Teneriffe	135
Mt. Athos, Greece	100

RIVERS OF THE WORLD.

NORTH AMERICA.

Names.	Length in m.	Names.	Length in m.
Missouri, to the Mississippi	3,100	Colorado (of the West)	1,000
Missouri, to the Gl. of Mexico	4,350	Ohio	948
Mississippi	3,160	Lewis, or Snake	900
Mackenzie's, fr. Slave Lake	900	Tennessee	800
Mackenzie's, fr. head of the Athabasca	2,440	Cumberland	600
St. Lawrence, from Lake Ontario	750	Appalachicola	600
St. Lawrence, from head of St. Louis R.	2,200	Mobile	550
Arkansas	2,170	Susquehannah	450
Rio Grande	1,800	James	450
Red	1,500	Sacramento	420
Platte, or Nebraska	1,500	Potomac	400
Nelson and Saskatchewan	1,600	Savannah	400
Columbia, or Oregon	1,200	Altamaha	400
		Connecticut	400
		Minnesota	375
		Roanoke	350
		Pedee	400
		Hudson	325
		Delaware	300
		Penobscot	300
		Mohawk	175
		Kennebec	150
		Genesee	145
		Merrimac	110

SOUTH AMERICA.

Names.	Length in m.	Names.	Length in m.
Amazon	4,000	Orinoco	1,500
La Plata, from the head of the Parana	2,250	St. Francisco	1,200
Madeira	1,800	Tocantins	1,500
Paraguay	1,500	Uruguay	1,000
		Purus	900
		Magdalena	900

EUROPE.

Names.	Length in m.	Names.	Length in m.
Volga	2,000	Elbe, Oder	550
Danube	1,725	Vistula	530
Dnieper	1,200	Weser	497
Don	995	Dniester	400
Rhine	900	Douro	450
Petchora	900	Guadiana	380
Ural	840	Dwina	350
Tagus	650	Po	350
Rhone	645	Thames	220
Loire	580		

ASIA.

Names.	Length in m.	Names.	Length in m.
Yang-tse Kiang	2,900	Euphrates	1,670
Lena	2,400	Indus	1,800
Yenisei	2,330	Ganges	1,500
Amoor	2,300	Brahmapootra	1,500
Obi	2,000	Amoo, or Oxus	1,300
Hoang Ho	2,600	Irrawaddy	1,200
Cambodia	2,000	Sihon	900
Irtysh	1,700	Meinam	800

AFRICA.

Names.	Length in m.	Names.	Length in m.
Nile	4,000	Senegal	1,200
Niger	2,700	Orange	1,000
Zambese	1,250	Gambia	700

The Grand Divisions of the Earth.

Grand Divisions.	Area.	Population.
Asia	16,415,758	753,962,433
Africa	11,556,650	192,000,000
North America	8,050,947	80,500,474
South America	6,554,125	27,092,800
Oceanica	4,288,025	30,348,251
Europe	3,800,557	301,753,419
Total	52,504,848	1,381,100,321

TENNESSEE.

Places represented on the map by numbers.

1 Troy,	15 Taylorsville,
2 Dyersburg,	16 Elizabethtown,
3 Paris,	17 Kingston,
4 Huntingdon,	18 Madisonville,
5 Dover,	19 Athens,
6 Waverly,	20 Decatur,
7 Ashland,	21 Dunlap,
8 Smithville,	22 McMinnville,
9 Livingston,	23 Manchester,
10 Jamestown,	24 Waynesboro,
11 Huntsville,	25 Decaturville,
12 Maynardville,	26 Savannah,
13 Tazewell,	27 Purdy,
14 Rogersville,	28 Bolivar.

NAMES OF CITIES AND TOWNS REPRESENTED ON THE MAPS BY NUMBERS.

[illegible]

POPULATION OF THE PRINCIPAL CITIES, TOWNS, AND VILLAGES IN THE UNITED STATES.

From the Census of 1870. * The star indicates the Population in 1875.

ALABAMA. — ARIZONA. — ARKANSAS. — CALIFORNIA. — CONNECTICUT. — COLORADO. — DELAWARE. — DIST. COLUMBIA. — FLORIDA. — GEORGIA. — ILLINOIS. — INDIANA. — IOWA. — KANSAS. — KENTUCKY. — LOUISIANA. — MAINE. — MARYLAND. — MASSACHUSETTS. — MICHIGAN. — MINNESOTA. — MISSISSIPPI. — MISSOURI. — NEBRASKA. — NEVADA. — NEW HAMPSHIRE. — NEW JERSEY. — NEW YORK. — NORTH CAROLINA. — OHIO. — OREGON. — PENNSYLVANIA. — RHODE ISLAND. — SOUTH CAROLINA. — TENNESSEE. — TEXAS. — UTAH. — VERMONT. — VIRGINIA. — WASHINGTON TER. — WEST VIRGINIA. — WISCONSIN.

[illegible]

HISTORICAL TABLE OF THE UNITED STATES.

States and Territories.	Where Settled.	By Whom Settled.	When.	When Admitted.	Area in Sq. Miles.	Pop. in 1870.
New Mexico	*Santa Fe*	*Spanish, &c.*	[illegible]		[illegible]	[illegible]
Florida	*St. Augustine*	*Spanish*	[illegible]	March 3, 1845	[illegible]	[illegible]
Virginia	*Jamestown*	*English*	[illegible]	June 26, 1788	[illegible]	[illegible]
New York	*Albany and New York*	*Dutch*	[illegible]	July 26, 1788	[illegible]	[illegible]
Massachusetts	*Plymouth*	*English*	[illegible]	Feb. 6, 1788	[illegible]	[illegible]
New Jersey	*Bergen*	*Dutch and Danes*	[illegible]	Dec. 18, 1787	[illegible]	[illegible]
New Hampshire	*Dover*	*English*	[illegible]	June 21, 1788	[illegible]	[illegible]
Maine	*York*	*English*	[illegible]	March 3, 1820	[illegible]	[illegible]
Connecticut	*Windsor*	*Emigrants from Mass.*	[illegible]	Jan. 9, 1788	[illegible]	[illegible]
Maryland	*St. Mary's*	*English*	[illegible]	April 28, 1788	[illegible]	[illegible]
Rhode Island	*Providence*	*Roger Williams*	[illegible]	May 29, 1790	[illegible]	[illegible]
Delaware	*Wilmington*	*Swedes and Finns*	[illegible]	Dec. 7, 1787	[illegible]	[illegible]
Pennsylvania	*Philadelphia*	*Swedes*	[illegible]	Dec. 12, 1787	[illegible]	[illegible]
North Carolina	*Albemarle*	*English*	[illegible]	Nov. 21, 1789	[illegible]	[illegible]
Wisconsin	*Green Bay*	*French*	[illegible]	May 29, 1848	[illegible]	[illegible]
South Carolina	*Port Royal*	*English*	[illegible]	May 23, 1788	[illegible]	[illegible]
Michigan	*Detroit*	*French*	[illegible]	Jan. 26, 1837	[illegible]	[illegible]
Illinois	*Kaskaskia*	*French*	[illegible]	Dec. 3, 1818	[illegible]	[illegible]
Arkansas	*Arkansas Post*	*French*	[illegible]	June 15, 1836	[illegible]	[illegible]
Indiana	*Vincennes*	*French*	[illegible]	Dec. 11, 1816	[illegible]	[illegible]
Texas	*Bexar*	*Spanish*	[illegible]	Dec. 29, 1845	[illegible]	[illegible]
Louisiana	*Iberville*	*French*	[illegible]	April 8, 1812	[illegible]	[illegible]
Alabama	*Mobile*	*French*	[illegible]	Dec. 14, 1819	[illegible]	[illegible]
Mississippi	*Natchez*	*French*	[illegible]	Dec. 10, 1817	[illegible]	[illegible]
Vermont	*Fort Dummer*	*Emigrants from Mass.*	[illegible]	March 4, 1791	[illegible]	[illegible]
Georgia	*Savannah*	*English*	[illegible]	Jan. 2, 1788	[illegible]	[illegible]
Missouri	*St. Genevieve*	*French*	[illegible]	Aug. 10, 1821	[illegible]	[illegible]
Tennessee	*Fort Loudon*	*Emigrants from N. C.*	[illegible]	June 1, 1796	[illegible]	[illegible]
California	*San Diego*	*Spanish*	[illegible]	Sept. 9, 1850	[illegible]	[illegible]
Kentucky	*Boonsboro*	*Daniel Boone*	[illegible]	June 1, 1792	[illegible]	[illegible]
Ohio	*Marietta*	*Em. from N. England*	[illegible]	Nov. 29, 1802	[illegible]	[illegible]
Oregon	*Astoria*	*Em. from N. England*	1811	Feb. 14, 1859	[illegible]	[illegible]
Washington					[illegible]	[illegible]
Iowa	*Burlington*	*Em. from N. England*	[illegible]	Dec. 28, 1846	[illegible]	[illegible]
Minnesota	*St. Paul*	*Em. from N. England*	[illegible]	May 11, 1858	[illegible]	[illegible]
Utah	*Salt Lake City*	*Mormon Emigrants*	1847		[illegible]	[illegible]
Nebraska				March 1, 1867	[illegible]	[illegible]
Kansas				Jan. 29, 1861	[illegible]	[illegible]
Colorado				Aug. 1, 1876	[illegible]	[illegible]
West Virginia				June 20, 1863	[illegible]	[illegible]
Nevada				Oct. 31, 1864	[illegible]	[illegible]

PRESIDENTS OF THE UNITED STATES.

Names.	Inaugurated.	Years served.
George Washington	1789	8
John Adams	1797	4
Thomas Jefferson	1801	8
James Madison	1809	8
James Monroe	1817	8
John Quincy Adams	1825	4
Andrew Jackson	1829	8
Martin Van Buren	1837	4
William H. Harrison	1841	1/12
John Tyler*	1841	3 11/12
James K. Polk	1845	4
Zachary Taylor	1849	1 1/3
Millard Fillmore*	1850	2 2/3
Franklin Pierce	1853	4
James Buchanan	1857	4
Abraham Lincoln	1861	[illegible]
Andrew Johnson*	1865	[illegible]
Ulysses S. Grant	1869	8
Rutherford B. Hayes	1877	

* *Elected Vice-President, and became President on the death of predecessor.*

THE FIRST THIRTEEN STATES.

New Hampshire,	New Jersey,	Virginia,
Massachusetts,	Pennsylvania,	North Carolina,
Rhode Island,	Delaware,	South Carolina,
Connecticut,	Maryland,	Georgia.
New York.		

ROUTES OF TRAVEL.

(See Map of the Western States.)

What cities do you pass on a voyage down the Ohio River? Down the Mississippi, on its right bank? Left bank?

To what cities can you sail on a voyage from Chicago to Cleveland? On what waters would you sail?

What cities would you pass in traveling by railroad from Cleveland to Chicago? St. Louis to Milwaukee? Alton to Columbus? Detroit to Dubuque?

Give the direction and distance from Chicago to Galena. Chicago to St. Louis. St. Louis to Cincinnati. Cincinnati to Detroit. Louisville to Michigan City.

If the State in which you reside be represented on this map, the following will be additional exercises:

State the direction from you of Cincinnati,—Chicago,—St. Louis,—Louisville,—Detroit,—Milwaukee,—Toledo. Point toward each.

Mention all the cities and towns in the northern part of your State. In the eastern,—southern,—western,—central part.

How many miles from you to the capital of your State? To its largest city?

Name all the places on the map within fifty miles of your residence.

What is the population of the largest cities in your State? (See "TABLES.")

Name the prominent places in both hemispheres which are in the same latitude as the city in or near which you reside. Draw a map of your State.

REVIEW.

CITIES AND TOWNS.

In what part of what State? On what waters? Which are capitals?

CINCINNATI, [illegible]	QUINCY,	KEOKUK,	LEXINGTON (MO.)
ST. LOUIS, [illegible]	DUBUQUE,	GRAND RAPIDS,	BLOOMINGTON,
CHICAGO, [illegible]	EVANSVILLE,	RACINE,	JEFFERSON CITY,
LOUISVILLE, 100	DAVENPORT,	JANESVILLE,	MINNEAPOLIS,
DETROIT,	ST. PAUL,	BURLINGTON,	WINONA,
MILWAUKEE,	FORT WAYNE,	KALAMAZOO',	FOND DU LAC, (WIS.)
CLEVELAND,	LAFAYETTE,	MADISON, (WIS.)	PRAIRIE DU CHIEN,
DAYTON,	ZANESVILLE,	HANNIBAL,	GRAND TRAVERSE,
INDIANAPOLIS,	LEXINGTON (KY.),	SPRINGFIELD,	FRANKFORT,
COLUMBUS,	ST. JOSEPH,	OSHKOSH,	LANSING,
COVINGTON,	SANDUSKY,	ALTON,	IOWA CITY,
PEORIA,	GALENA,	ROCK ISLAND,	NEW ALBANY,
TOLEDO,	MADISON (IND.),	MUSCATINE,	KANSAS,
TERRE HAUTE, (ter'reh-hote')	CHILLICOTHE, (chil-le-koth'ee)	DES MOINES, (de moin')	COUNCIL BLUFFS.

RIVERS.

Where do they rise? Between or through what States do they flow? Into what waters do they flow?

MISSISSIPPI? 40*	MIAMI?	SANDUSKY?	WHITE?
OHIO? 9½	MUSKINGUM?	GRAND (MICH.)?	BLACK?
TENNESSEE? 8	KANKAKEE?	GRAND (MO.)?	GREEN?
CUMBERLAND? 6	MINNESOTA?	IOWA?	RED?
WISCONSIN?	KENTUCKY?	MAUMEE?	ROCK?
WABASH?	KANSAS?	RAINY LAKE?	WOLF?
DES MOINES?	NEBRASKA?	RED CEDAR?	OSAGE?
DETROIT?	DAKOTA?	BIG SANDY?	LICKING?
CHIPPEWA?	KASKASKIA?	BIG SIOUX?	ST. CLAIR?
ILLINOIS? (il-li-noy')	SANGAMON? (sang'ga-mon)	LITTLE SIOUX? (soo)	ST. CROIX? (sent kroy')

LAKES.

By what land are they enclosed? What are their outlets?

SUPERIOR? 350*	WINNEBAGO?	ERIE? 240	PEPIN?
MICHIGAN? 300	L. OF THE WOODS?	ITASCA?	RAINY?
HURON? 260	BIG STONE?	RED?	ST. CLAIR?

* Numbers in list of Cities and Towns show population in *thousands*; in that of Rivers, the length in *hundreds of miles*; in that of Lakes, the whole length in miles.

MAP DRAWING.

THE SCALE.

It must be evident to all thinking teachers that maps should be constructed on the *same scale* by all the scholars in the school. By the use of a single measure, the State or Country is accurately drawn, its area, compared with that of other States and Countries is at once seen, and comparative size fixed in the mind. It is certainly very important for pupils to have accurate ideas of the comparative sizes of the different States of the Union and the countries of the world.

Each measure on the scale shown below always represents *two hundred miles*. *One-half* represents the distance of *one hundred miles*; *one-fourth* of *fifty miles*, and *one-eighth*, of *twenty-five miles*. The other subdivisions are shown on the scale.

EXERCISES ON THE USE OF THE SCALE

should be continued until the scholars are quick and accurate in its use. *Squares* as well as lines should be drawn, representing each of the divisions of this measure.

SCALE FOR DRAWING ALL THE STATES.

If a distance is a little longer than is represented by the scale, it is indicated by the sign plus (+), and if it is a little shorter, it is indicated by the sign minus (−). Any distance, however small, may be represented by still more minutely subdividing the measure, but for all practical purposes the divisions indicated are sufficiently minute.

DRAWING A MAP.

The teacher is now prepared to commence the work of constructing a map. Let some State of a regular shape be first chosen, as Pennsylvania. The southern boundary is *one and one-third*, the northern boundary the same, and the width *three-fourths*. The notch in the N. W. corner of the State is a little less than *one-fourth*. This is a part of *one and one-third*, the extreme length of the northern boundary line. The Ohio River crosses the western boundary line of the State *one-fourth* north of the S. W. corner.

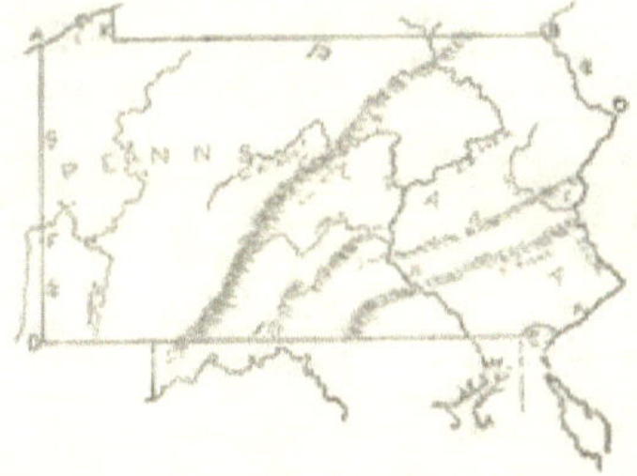

As soon as Pennsylvania is drawn it should be made a basis upon which New York and the New England States are constructed. Notice how easily this can be done. From the N. E. corner of Pennsylvania *one-half east* reaches the S. W. corner of Massachusetts. This point is a centre from which *one-half* a little west of south, reaches Brooklyn; *one-half east*, the N. W. corner of Rhode Island; *one-half north*, the southern point of Lake Champlain, and *two halves* or *one measure*, the N. E. corner of the State of New York.

The various distances are all marked on the construction lines, so that by noticing the outline maps that follow, no difficulty will be experienced in constructing all the States.

MAP-DRAWING ON A UNIFORM SCALE.

STATES SHOULD BE GROUPED.

Pennsylvania is the base for New York, Virginia, Delaware, and Maryland. The New England States can more easily be drawn together than separately. By this means their comparative size is fixed in the mind.

ENLARGED MAPS

of single States or groups of States can easily be drawn by taking a longer unit of measure than is given in this book. Blackboard maps as large as may be wished can be executed by using a blackboard scale.

ASSOCIATION OF SIMILAR DISTANCES

should accompany every exercise in mapdrawing.

Such a remarkable similarity in distance exists in the boundary lines of the United States and the Countries of Europe, that by associating their lengths, no difficulty will be experienced in remembering them.

The One-half Measures around the S. W. corner of Massachusetts have already been mentioned. Among many other coincidences, it may be noticed that the southern boundary of Iowa, the width of Iowa, the extreme width of Illinois, and the width of Kansas, are all *one measure*.

The length of the northern boundary of Iowa is the same as the extreme length and breadth of Missouri, and these are the same as the length of the peninsula of Michigan.

These are only a few of the many remarkable coincidences found to exist in the United States. By associating similar distances no difficulty will be found in teaching pupils how long and broad the States are.

We now shall show

HOW TIME MAY BE SAVED IN TEACHING TOPICAL GEOGRAPHY BY MEANS OF MAP DRAWING.

Pupils should come to the class with *paper, a scale* and *a pencil*. *First* they are instructed in the use of the scale; *then* they are required to draw the outlines of the States or Countries previously given. This may be done with the map drawing book open before them, until they have learned to draw the lesson independently of the book.

THE ORDER OF RECITATION.

After a sufficient time has been spent in drawing outline maps, in accordance with the measurements given in the book, then the pupils may be required to bring to the class-room an outline map *previously* prepared. This will save time.

Suppose the lesson is the State of New York. Each pupil is prepared with an outline map of this State, to be filled up at the dictation of the teacher. When all are ready the pupils are given a sufficient time in which to draw the *mountains*, then the *rivers*, and after this the *towns* are located, noticing that the mountains control the courses of rivers, and the rivers the locations of towns. At the close of the recitation these papers are collected, corrected and returned. By pursuing this course each pupil has recited, and in such a way as to give the best possible proof of his knowledge of the geography of the State. It takes only a short time, and is a thorough test.

All class exercises in map-drawing should be divided into two distinct parts:

I. Exercises in drawing outline maps and in distances;

II. Exercises in completing outline maps previously drawn.

It is not necessary to continue these general directions further, as the figures near the measurement lines give sufficient information to enable any pupil of ordinary capacity to construct any map in this book. It is suggested that the following order be observed in the construction of the map.

I. The Outlines. *II. Mountains and Surface Elevations.*
III. Rivers and Lakes. *IV. Cities and Towns.*
V.—Canals and Railways.

Navigable Rivers may be drawn with double lines.

Coast Lines need not be shaded, but if any prefer to finish the map in this way, five or six lines may be drawn, each conforming as nearly as possible to the direction of the coast. The last ones should be finer and further apart than the first.

Mountains may be drawn as on the maps in this book. It should be noticed that very high elevations should not be drawn in the same manner as lower elevations or single peaks. These different methods are represented in the maps of this book. Maps may be made very distinct by first sketching the whole work with a lead pencil, and then tracing it in ink with a fine steel pen.

COLORING.

A map can be colored only when drawn in lead pencil or with india ink. The object of this may be to more clearly distinguish the *political divisions* or the physical features. In the former case several colors are necessary, in the latter, only two or three. Let the paints be of good quality, dissolved in water, and made to flow like inks, and then spread upon the paper very much diluted.

Yellow Ink may be made by dissolving *gamboge* in pure rain-water.

Blue Ink.—Dissolve an ounce of *Prussian Blue* in one pint of water, in which one ounce of *Oxalic Acid* has been dissolved. Add a small quantity of *Gum Arabic*.

Red Ink.—*Carmine* dissolved in *liquid Ammonia* is the usual method of making red ink. After it is dissolved dilute it with pure rain-water.

Green Ink.—Mix blue and yellow inks. *Purple Ink.*—Blue and red.

Two or three coats should be applied with a medium sized camel-hair or sable-hair brush. If two colors only are used, red and green, or yellow and blue should be selected. After the map is colored the boundary lines may be rendered more distinct by tracing them with good carmine ink, using a fine brush. With these directions, even pupils possessing only ordinary ability in drawing may be able to construct and finish very beautiful and correct maps.

HOW TO CONSTRUCT THE CONTINENTS.

The scale, representing the distance of two hundred miles, described above, is only used in constructing the United States and other *subdivisions* of the Continents.

For constructing the Continents a scale, each division of which represents *six hundred miles*, is used.

SCALE FOR DRAWING ALL OF THE CONTINENTS.

THE SIX HUNDRED MILE SCALE.

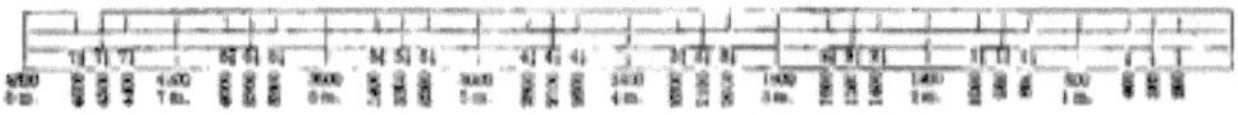

4800 8 m. … 4200 7 m. … 3600 6 m. … 3000 5 m. … 2400 4 m. … 1800 3 m. … 1200 2 m. … 600 1 m.

Larger maps may be easily drawn by increasing the length of the measure.

Note.—An excellent ruler, marked with these two map-drawing scales on one side and inches on the other, can be obtained from A. S. Barnes & Co., New York, or any bookseller in the country. It is as cheap as any common measure, and is useful for general desk purposes. A longer scale for the blackboard is also furnished.

MAP DRAWING ON A UNIFORM SCALE.

MAINE.

Make a scale on a slip of stiff paper, and by it draw maps of all the States, as explained on a previous page.

Draw no lines except boundaries.

Commence at A, draw *Passamaquoddy Bay*, and locate *Eastport*. Measure west, on the 45° of latitude, and measure to B; thence [illegible] m. to L, the northeast corner of *N. H.*, and [illegible] m. to H, the northeast corner of *Vt.*

From H measure [illegible] m. to the *Salmon Falls River* at F, and draw the western boundary. From F measure [illegible] m. to G, and draw the *Salmon Falls River*, a part of the *Merrimac River*, and the coast line.

From the point C, a little less than [illegible] m. east of B, measure [illegible] m. to D, the most northern point of *Me.*; thence [illegible] m. towards H to K, and complete the northwestern boundary of the State.

From A measure [illegible] m. toward D, to E, and draw *Grand Lake* and *St. Croix River*. From E measure north to I [illegible] m., and draw the eastern boundary line and the *River St. John*.

MASSACHUSETTS, CONNECTICUT, AND RHODE ISLAND.

Begin at H, and measure [illegible] m. to G and [illegible] m. to K, and draw the northern boundary of the State, the *Merrimac River* and *Cape Ann*.

Draw the western boundary, [illegible] m. from G to P. From P, measure [illegible] m. a little west of south to R, and draw the western boundary of *Connecticut* and the western end of *Long Island*. From P, measure [illegible] m. to N, [illegible] m. from N to M, [illegible] m. from M to L, [illegible] m. from L to O, and [illegible] m. from O south to S. Draw the southern boundary of *Connecticut* and *Rhode Island*; *Plymouth Bay*, *Cape Cod*, *Cape Cod Bay*, and the eastern coast of *Massachusetts*; *Martha's Vineyard* and *Nantucket*.

Draw the boundary line between *Connecticut* and *Rhode Island*, [illegible] m., a little west of N.

Draw the northern shore of *Long Island Sound*, *Narragansett Bay*, and *Buzzard's Bay*.

Complete these States by adding the mountains, rivers, principal towns, and railroads.

Draw *Long Island*, observing that its eastern extremity extends to a point directly south of the eastern boundary of *Connecticut*.

Many questions may now be asked like the following:

What is the length, in miles, of the western boundary of Rhode Island? Of the western boundary of Massachusetts?

If the teacher wish, the pupils may now draw the six Eastern States together.

Observe how many distances, in these States, are exactly the same.

NEW HAMPSHIRE AND VERMONT.

The measurements for the eastern boundary of New Hampshire are the same as those for the western boundary of Maine. Make a scale like the one given at the foot of page 27.

Begin at A, measure [illegible] m. north to B, [illegible] m. north to J, [illegible] m. from J to L, and [illegible] m. east from L to K. Complete the eastern boundary of the State, drawing *Salmon Falls River*, the *Atlantic Coast*, and *Cape Ann*. Measure [illegible] m. from L to H, and [illegible] m. from H to C. Draw *Merrimac River*, and complete the southern boundary of the States.

From A, measure [illegible] m. west to C, [illegible] m. from C to E, [illegible] m. from E south to F, and [illegible] m. from F to G. Draw *Lake Champlain*, *Connecticut River*, the *Green* and *White Mountains*.

Complete the drawing by adding, from the large map, the rivers,—the bays,—the capes,—the cities and towns,—the railroads.

NOTE.—Here the drawings may be examined by the teacher, or the pupils may draw the map on the blackboard, each doing a part.

NEW YORK

Begin at E, measure [illegible] m. south to F, [illegible] m. from F to G, [illegible] m. from G to P, and [illegible] m. a little west of south to R. Draw *Lake Champlain*, the eastern boundary of the State, *Long Island*, *Long Island Sound*, *Staten Island*, *Sandy Hook*, and the southern extremity of *Hudson's River*. Locate *New York*, *Brooklyn*, and *Jersey City*.

From P measure [illegible] m. west to B. From B measure [illegible] m. toward R and draw a part of *Delaware River* and the northern boundary of *New Jersey* [illegible] m.

From B measure a little more than [illegible] m. west to K, [illegible] m. north from K to L, and [illegible] m. east from L to M, and draw the northern boundary of *Pennsylvania* and the eastern extremity of *Lake Erie*. Locate *Buffalo* and *Niagara Falls*.

From L measure [illegible] m. north to N, and [illegible] m. east from N. Draw *Lake Ontario*, observing that it is [illegible] m. wide.

From E measure [illegible] m. west to D, and [illegible] m. from D toward K to S, and draw *St. Lawrence River*. Complete the State by marking the mountains, rivers, cities, railroads, and the Erie Canal.

MAP DRAWING ON A UNIFORM SCALE.

PENNSYLVANIA AND NEW JERSEY. | OHIO, INDIANA, AND KENTUCKY.

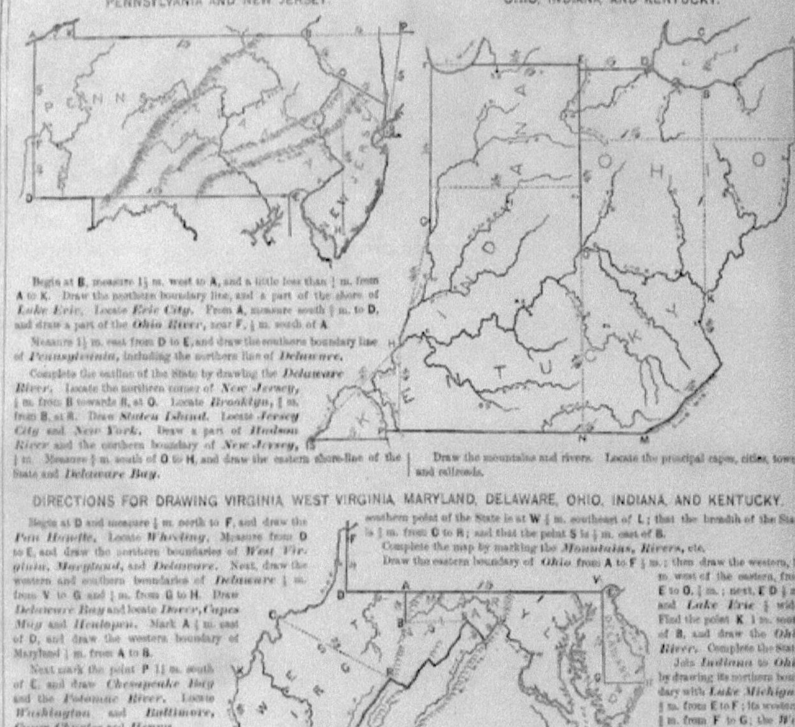

Begin at **B**, measure 1½ m. west to **A**, and a little less than [illegible] m. from **A** to **K**. Draw the northern boundary line, and a part of the shore of ***Lake Erie***. Locate ***Erie City***. From **A**, measure south [illegible] m. to **D**, and draw a part of the ***Ohio River***, near **F**, [illegible] m. south of **A**.

Measure 1½ m. east from **D** to **E**, and draw the southern boundary line of ***Pennsylvania***, including the northern line of ***Delaware***.

Complete the outline of the State by drawing the ***Delaware River***. Locate the northern corner of ***New Jersey***, [illegible] m. from **B** towards **R**, at **O**. Locate ***Brooklyn***, [illegible] m. from **B**, at **R**. Draw ***Staten Island***. Locate ***Jersey City*** and ***New York***. Draw a part of ***Hudson River*** and the northern boundary of ***New Jersey***, [illegible] m. Measure [illegible] m. south of **O** to **H**, and draw the eastern shore-line of the State and ***Delaware Bay***.

Draw the mountains and rivers. Locate the principal capes, cities, towns, and railroads.

DIRECTIONS FOR DRAWING VIRGINIA, WEST VIRGINIA, MARYLAND, DELAWARE, OHIO, INDIANA, AND KENTUCKY.

Begin at **D** and measure [illegible] m. north to **F**, and draw the ***Pan Handle***. Locate ***Wheeling***. Measure from **D** to **E**, and draw the northern boundaries of ***West Virginia***, ***Maryland***, and ***Delaware***. Next, draw the western and southern boundaries of ***Delaware*** [illegible] m. from **V** to **G** and [illegible] m. from **G** to **H**. Draw ***Delaware Bay*** and locate ***Dover***, ***Capes May*** and ***Henlopen***. Mark **A** [illegible] m. east of **D**, and draw the western boundary of Maryland [illegible] m. from **A** to **B**.

Next mark the point **P** 1½ m. south of **E**, and draw ***Chesapeake Bay*** and the ***Potomac River***. Locate ***Washington*** and ***Baltimore***, ***Capes Charles*** and ***Henry***.

Draw the southern boundary line of ***Virginia*** 1½ m. from **P** to **N**, and [illegible] m. from **N** to **M**. From **M** measure [illegible] m. northeast to **L**, and draw the ***Cumberland Mountains***. Measure north [illegible] m. from **L** to **K**, and draw the ***Big Sandy River***. Draw the ***Ohio River*** from **F** to **K**.

Complete the eastern boundary of ***West Virginia***, observing that the southern point of the State is at **W** [illegible] m. southeast of **L**; that the breadth of the State is [illegible] m. from **C** to **R**; and that the point **S** is [illegible] m. east of **B**.

Complete the map by marking the ***Mountains***, ***Rivers***, etc.

Draw the eastern boundary of ***Ohio*** from **A** to **F** [illegible] m.; then draw the western, 1½ m. west of the eastern, from **E** to **O**, [illegible] m.; next, **E D** [illegible] m. and ***Lake Erie*** [illegible] wide. Find the point **K** 1 m. south of **B**, and draw the ***Ohio River***. Complete the State.

Join ***Indiana*** to ***Ohio*** by drawing its northern boundary with ***Lake Michigan*** [illegible] m. from **E** to **F**; its western, [illegible] m. from **F** to **G**; the ***Wabash River*** [illegible] m. from **G** to **H**; and the ***Ohio River*** from **O** to **H**.

Draw ***Kentucky*** by measuring [illegible] m. from **C** south to **N**. Mark the southern boundary 1½ m. from **M** to **P**, the ***Tennessee River***; and [illegible] m. from **P** to **S**, the ***Mississippi River***. Locate the principal rivers, mountains, etc., as in the other maps.

MAP DRAWING SHOULD BE SYSTEMATIC.

If one State or Continent is drawn on a definite plan, the same course should be pursued with all. States should be so drawn as to be joined to other States, and Continents to other Continents. This is not possible by some methods of drawing before the public. Vermont is drawn on one scale, and Massachusetts on another, while Connecticut is drawn on still another, and New York on quite another; so that it is impossible to join all of these neighboring States in studying their common physical features.

PRINCIPLES UNDERLYING THE SUBJECT.

1. Maps should be drawn in accordance with a definite unit of measurement.
2. Actual distances should be learned.
3. States should be studied in groups, and these should be united as the lessons advance, and thus form entire sections.

ORDER OF DRAWING A MAP.

I. Measurement Outlines.
II. Boundary Lines.
III. Mountains and Surface Elevations.
IV. Rivers. Inland Lakes.
V. Cities and Towns.
VI. Railroads and Canals.
VII. Write in each State or Continent its principal products and the leading occupations of its inhabitants.

MAP DRAWING ON A UNIFORM SCALE.

WISCONSIN AND MINNESOTA

Begin at F, and measure ¼ m. north, and draw their southern boundaries, **A B C D**; thence, 1¾ north to **H**. Fix the points **E, G, L, M,** and **N**, as indicated, and complete the States.

MICHIGAN.

Form the square **C N I F, each side 1¼ m.** long, and subdivide into four squares. **Draw** the *Strait of Mackinaw*, *Lakes* ***Michigan***, *Huron*, *St. Clair*, and *Erie*; **then** the southern boundary. At S, ¼ m. north **of** the *Strait of Mackinaw*, fix the southeastern extremity of *Lake Superior*; thence 1¾ west to the western extremity, K, and draw *Lake Superior*, noticing that the northern coast at R is north of the west coast of *Lake Michigan*.

ILLINOIS, IOWA AND MISSOURI.

Draw the eastern boundary line of ***Illinois***, the same as the western boundary line of ***Indiana***, ¾ m. from *Lake Michigan* to ***Wabash River***. From **F**, measure ¼ m. north **and** ¼ m. west to **A**, and draw the southern coast-line of *Lake Michigan*. Locate *Chicago*.

Mark the northern boundary ¾ m. from **A** to **B**. Locate *Cairo* 1¾ m. south of the line **A B**. Measure the extreme breadth of the State 1¼ m. on the line **V U**, and draw the ***Mississippi***, ***Ohio***, and ***Wabash Rivers***. Complete drawing of State.

Next draw ***Iowa***, commencing with its northern boundary 1¾ m. from **C** to **D**, ¼ m. north of the northern boundary of ***Illinois***, and ½ m. west. From **Y**, ¼ m. west of **C**, measure 1 m. south to **E**, and draw the southern boundary 1¾ m. to **K**. Draw the ***Des Moines River***. The eastern bend of the river is ¾ m. east of the central line **YE**. Complete the drawing according to the measurements.

Complete ***Missouri***, by commencing at **K** and measuring ¼ m. east and ¼ m. south to ***Kansas City***. Mark its southern boundary 1¾ m. south of **its northern, 1¼ m.** in length from **N** to **T**, and ¾ m. **from T to S**.

MAP DRAWING ON A UNIFORM SCALE.

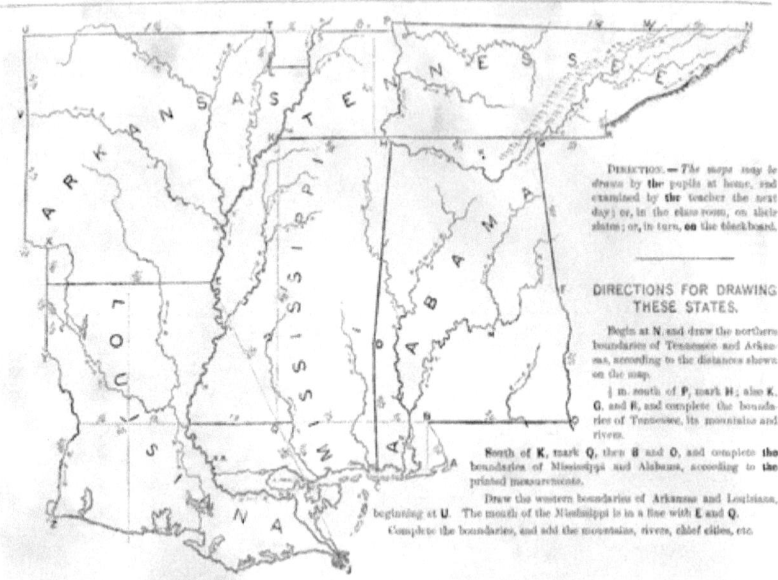

DIRECTION.—*The maps may be drawn by* **the** *pupils at home, and examined by* **the** *teacher the next day; or, in the class room, on their slates; or, in turn,* **on** *the blackboard.*

DIRECTIONS FOR DRAWING THESE STATES.

Begin at **N** and draw the northern boundaries of Tennessee and Arkansas, according to the distances shown on the map.

½ m. south of **P**, mark **H**; also **K**, **G**, and **R**, and complete the boundaries of Tennessee, its mountains and rivers.

South of **K**, mark **Q**, then **B** and **O**, and complete **the** boundaries of Mississippi and Alabama, according to **the** printed measurements.

Draw the western boundaries of **Arkansas and Louisiana**, beginning at **U**. The mouth of the Mississippi is in a line with **E and Q**.

Complete the boundaries, and add the mountains, rivers, *chief cities*, etc.

OUTLINE OF A TOPICAL RECITATION.

First of all, DRAW THE MAP, then tell in order:

I. POSITION ON THE GLOBE. Let this be given exactly. Latitude and Longitude.

II. BOUNDARY, MEASUREMENT LINES, as learned from the construction of the map, general shape, number of square miles it contains, character of boundary lines, as mountain ranges, rivers, straits, bays, peninsulas, isthmuses.

III. SURFACE ELEVATION; mountain ranges, plateaux, slopes, single peaks, inland waters, average elevation above the ocean. Name each river, stating its source, direction, length, and where it empties.

IV. CLIMATE. State causes regulating it, as affected by latitude and longitude, altitude, nearness to, or remoteness from, large bodies of water, and high mountain ranges.

V. VEGETATION; natural, cultivated, character of soil.

VI. ANIMALS; domestic, wild.

VII. INHABITANTS; original character of, present character of.

VIII. GOVERNMENT; how laws are made, the name and character of the government, principal officers, how elected.

IX. INTERNAL IMPROVEMENTS. Name the character, extent, and cost of each, also name the manufactures, industries, exports, imports.

X. PRINCIPAL TOWNS; where located, size, and trade of each.

XI. HISTORY; when settled, when, and by whom. WARS. NOTED MEN.

Special formulas can easily be given for describing mountains, rivers, oceans, bays, gulfs, and straits.

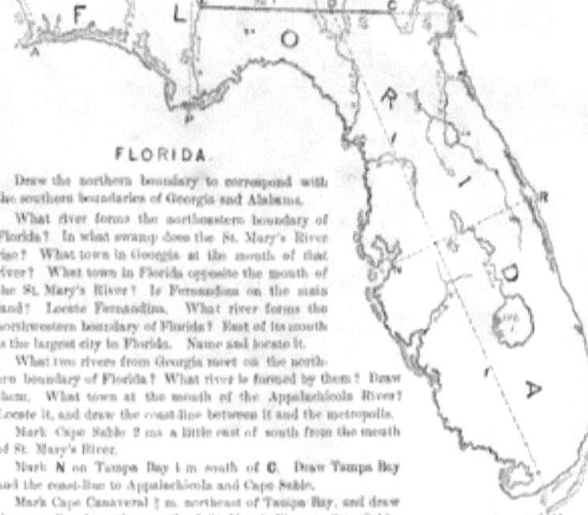

FLORIDA.

Draw the northern boundary to correspond with the southern boundaries of Georgia and Alabama.

What river forms the northeastern boundary of Florida? In what swamp does the St. Mary's River rise? What town in Georgia at the mouth of that river? What town in Florida opposite the mouth of the St. Mary's River? Is Fernandina on the main land? Locate Fernandina. What river forms the northwestern boundary of Florida? East of its mouth is the largest city in Florida. Name and locate it.

What two rivers from Georgia meet on the northern boundary of Florida? What river is formed by them? Draw them. What town at the mouth of the Appalachicola River? Locate it, and draw the coast-line between it and the metropolis.

Mark Cape Sable 2 ms a little east of south from the mouth of St. Mary's River.

Mark **N** on Tampa Bay 1 m south of **C**. Draw Tampa Bay and the coast-line to Appalachicola and Cape Sable.

Mark Cape Canaveral ½ m. northeast of Tampa Bay, and draw the coast-line from the mouth of St. Mary's River to Cape Sable. Complete the map.

MAP DRAWING ON A UNIFORM SCALE.

NORTH CAROLINA, SOUTH CAROLINA, AND GEORGIA

Draw the northern boundary of *North Carolina*, 1¾ measures in length. Mark the mouth of *St. Mary's River*, at S, 2 ms. south of N. From P, 1½ ms. toward S, mark K, the most southern point of *North Carolina*, and draw its coast line, with its sounds and capes; also their names.

Mark O, 1 m. west of P; L, ¾ m. south of O; H, 1 m. west of L; and R, ½ m. west of H. Draw the eastern and northern boundaries of South Carolina and the western boundary of North Carolina.

From A, the middle of the northern boundary of *South Carolina*, measure south 1 m. to B, and draw the *Savannah River*; also the coast-line of *South Carolina* and *Georgia*.

Mark G, ½ m. west of H; E, the junction of the *Flint* and *Chattahoochee Rivers*, 1½ m. south of G, and 1 m. west of S, and complete the boundaries. Mark the capes, cities, etc.

SUGGESTIONS.

For the purpose of presenting the entire map of the United States in one view, the scale has been reduced so as to show the principal measurements on one page.

On all of the other map-drawing maps in this book two inches represents one m., or two hundred miles. On the map below, one-half an inch is one m.

A scale may be prepared, suited to the size of the paper or board; but when this measure is once established in the school-room, it should never be changed. All other scales should conform to it.

In no other way can comparative size and area be taught. In drawing the United States as a whole, it should be remembered that all meridian lines point towards the pole.

THE UNITED STATES.

MAP DRAWING ON A UNIFORM SCALE.

SOUTH AMERICA.

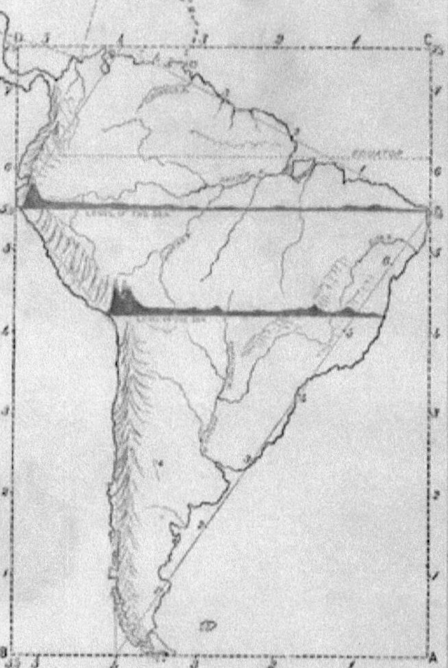

Commence at **A**, and measure 5½ ms. north, and mark ***Cape St. Roque.*** Measure to **C**, 7½ ms.; thence west to **D**, 5½ ms. At 4 ms. from **C**, mark ***Cape Gallinas*** and ***Lake Maracaybo.*** From 5½, at ***Cape St. Roque,*** toward 4 on the line **C D**, mark the points, 1, near the mouth of the ***Amazon River;*** 2, opposite ***Georgetown;*** and, 3, near ***Caracas.*** Complete the coast-line.

From **A**, toward the west, mark the points 4 and 5½ at **B**. From 4, west of **A**, toward ***Cape St. Roque,*** mark 1, near the ***Gulf of St. George;*** 2, opposite ***St. Matthias' Bay;*** 3, near the mouth of the ***Rio de La Plata;*** and 6, opposite the ***Bay of All Saints.*** Complete the coast-line.

From 4, west of **A**, measure 4 ms. north, and draw the coast south to ***Terra del Fuego*** and ***Cape Horn.***

North of **B**, mark the points 5½, 6, 7, and draw the ***Gulf of Darien, Isthmus of Panama,*** and ***Cape Blanco.*** Complete the drawing by marking the mountains, rivers, countries, bays, gulfs, capes, cities, etc., writing the *full* name of each outside the map.

With these directions, no pupil old enough to study Geography will have difficulty in drawing an accurate map of South America, giving its entire length and breadth, as well as the lengths of the coast-lines.

The internal construction can be easily drawn by referring to the maps in the Geography.

REMARKS.

ALL of the continents in this system of map drawing are drawn on the same scale, thus representing to the eye their comparative areas. They are also drawn on the same plan or system; so that, as soon as the method of constructing South America is learned, the pupil will have no difficulty in drawing North America and the other continental divisions.

The unit of measure is six hundred miles.

It will be seen that a fundamental principle in this system of map drawing is, that maps should be drawn to a single unit of measure. The great advantages of this will be at once apparent to teachers. Comparative size is fixed in the mind, and the lengths and breadths of the States and Continents easily learned. Maps are made to suit the maps, and not the maps to suit the pages.

But it may be convenient to enlarge Europe, and contract Asia and Africa. A unit of measure can easily be taken longer or shorter than the one here given, and the same method pursued as here indicated. Thus a map can be drawn as large or small as may be desired.

The continental unit assumed in this book is convenient for use at the desk.

All pupils should have the same unit of measure.

The maps can thus be easily compared, and in a short time the comparative sizes of the States and Continents will be learned.

SCALE FOR DRAWING ALL OF THE CONTINENTS.

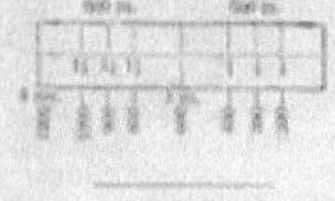

QUESTIONS.

Many questions will suggest themselves to the teacher. Among them we would suggest the following:

What is the extreme length of South America compared with North America? How does the extreme breadth of the United States compare with the breadth of Europe? What is the extreme length of the Gulf of Mexico? Its breadth? How far is it from Yucatan to Appalachee Bay? What is the length of Cuba? Its breadth? How far is Newfoundland from Hayti? How far is Hayti from South America? How far is Hayti from Greenland? What is the distance from Cape Catoche to Cape Sable? From Cape Sable to the Straits of Bellisle? How far is Queen Charlotte's Island north of San Francisco Bay?

MAP DRAWING SAVES TIME.

It is certain that much time now spent in learning local geography is lost because pupils do not gain a distinct mental view of the world on which they live. This can, to a good degree, be remedied by map drawing, and a much more permanent impression made, and thus much time now spent in reciting names can be saved.

At least *one-half* the time now spent in studying geography can be saved and much more accomplished. In order to permanently remember the location of a place, its name must be associated with its position on our earth. This can be much more easily accomplished by the aid of map drawing than by any other means. A teacher of large experience recently remarked that, in his opinion, "by means of map drawing *twice* as much could be learned in the same time with *ten times* the probability of its being remembered."

HOW TO CONDUCT A RECITATION.

Suppose the map of the *State* of New York is to be recited. The pupils are expected to know the principal characteristics of its local geography. Its *Mountains, Islands, Bays, Sounds, Straits, Rivers, Lakes, Falls, Cities, Towns, Railroads,* and *Canals* are to be recited in such a manner as to give the best evidence that each pupil knows their exact location. By the old method each pupil recites orally, with no delineation on the board, slate, or paper.

There is no certainty that all the pupils have obtained the entire lesson, as no one can recite the whole of it, and it consumes much time. Many names are learned, but accurate geographical knowledge has not been promoted.

A BETTER WAY.

Let each pupil, either at the board or on slate or paper, draw an outline map of the State. It need not take over two minutes.

Next, draw from the large map, in the following order, the mountains,—the rivers,—the bays,—the capes,—the cities and towns—mark those only which appear on the large map in *black* letters; then mark the railroads. In drawing a map of your own State, mark all the cities and towns.

Then, on the side of the map, let their names be written, corresponding to the numbers, as above.

The work can now be easily examined, as in an exercise in written spelling.

The entire work here mentioned need not take over *fifteen minutes*. *Every scholar has recited,* and in such a manner as to give the very best evidence of his knowledge or want of knowledge of the lesson.

If there is time after this work is inspected and corrected, then the usual oral recitation can proceed by requiring the pupil reciting to point to the *Mountains* on the outline drawn, while he is telling in what part of the State they are, and in what direction the ranges extend; or the *Islands,* where they are, by what waters surrounded, and so on until the map has been recited.

No names should be written on the face of the map drawn. In a short time pupils will obtain great skill in doing this work, and teachers will find it pleasant, expeditious, easy, and thorough.

www.ingramcontent.com/pod-product-compliance
Lightning Source LLC
Chambersburg PA
CBHW022133160426
43197CB00009B/1261

9783337132767